QH 491 .E33 Cop.1

Ede, D
An introduction to developmental biology

CHICAGO PUBLIC LIBRARY
HAROLD WASHINGTON LIBRARY CENTER

R0016483575

DATE

QH 491 .E33 cop.1

FORM 125 M

Business/Science/Technology Division

The Chicago Public Library

Received_____ MAY 4 1979 _____

© THE BAKER & TAYLOR CO.

An Introduction to Developmental Biology

TERTIARY LEVEL BIOLOGY

A series covering selected areas of biology at advanced undergraduate level. While designed specifically for course options at this level within Universities and Polytechnics, the series will be of great value to specialists and research workers in other fields who require a knowledge of the essentials of a subject

Titles in the series:

Experimentation in Biology	Ridgman
Methods in Experimental Biology	Ralph
Visceral Muscle	Huddart and Hunt
Biological Membranes	Harrison and Lunt
Comparative Immunobiology	Manning and Turner
Water and Plants	Meidner and Sheriff
Biology of Nematodes	Croll and Matthews
An Introduction to Biological Rhythms	Saunders
Biology of Ageing	Lamb
Biology of Reproduction	Hogarth
An Introduction to Marine Science	Meadows and Campbell
Biology of Fresh Waters	Maitland
An Introduction to Developmental Biology	Ede

TERTIARY LEVEL BIOLOGY

An Introduction to Developmental Biology

Donald A. Ede, B. Sc., M.S., Ph.D.
Reader in Zoology
University of Glasgow

A HALSTED PRESS BOOK

John Wiley and Sons
New York—Toronto

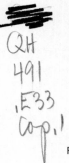

Blackie & Son Limited
Bishopbriggs
Glasgow G64 2NZ

450 Edgware Road
London W2 1EG

Published in the U.S.A. and Canada by
Halsted Press,
a Division of John Wiley and Sons Inc.,
New York

© 1978 D. A. Ede
First published 1978

All rights reserved.
No part of this publication may be reproduced,
stored in a retrieval system, or transmitted,
in any form or by any means,
electronic, mechanical, recording or otherwise,
without prior permission of the Publishers

Library of Congress Cataloging in Publication Data
Ede, D A
An introduction to developmental biology

(Tertiary level biology)
"A Halsted Press book."
Bibliography: p.
Includes index.
1. Developmental biology. I. Title.
QH491.E33   591'.3   78-16359
ISBN 0-470-26469-1

Filmset and printed in Great Britain by
Thomson Litho Ltd, East Kilbride, Scotland

Preface

SOME OF THE PROFOUNDEST BIOLOGICAL PROBLEMS ARE LOCATED IN THE FIELD of developmental biology, and an ever-increasing proportion of the most exciting work in cell biology, molecular biology and genetics is orientated towards it, especially towards the central problem of how by interaction at these various levels—molecule, gene and cell—the whole organism is produced. There is an exhilarating sense that from this new synthesis important concepts are emerging concerning how complex, differentiated and regulating structures are produced from simple cellular beginnings, and the most important thing an introductory book or course can do is to convey that sense and give it substance through observations, experiments and ideas representative of those which are coming from laboratories all over the world.

This book aims to provide the basis of a course in animal developmental biology suitable for advanced undergraduates in biology and for graduates coming from other fields—assuming no previous knowledge, keeping ideas well afloat without submerging them in a sea of facts, and covering all of the main aspects of the subject in a way which leads directly into the research literature. To this end there is a "further reading" list for each chapter in which one or more books or review articles are recommended for a more detailed overview of each topic, followed by specific references in which the original work described may be followed up in depth.

Sufficient descriptive material is woven into the text and illustrations to make the essential features of morphogenesis in the major vertebrate and invertebrate groups clear, but the main emphasis is upon experimental work which throws light on the underlying mechanisms of development. Its cellular basis is stressed, but there is a shift from concentration on cell differentiation in itself to the subject of its global control, i.e. pattern formation. And since genetic control plays so fundamental a part, considerable space is given to development in mammals and insects, especially to the mouse and *Drosophila*, in which species so much is available in the way of genetic material and techniques that genetics and developmental biology begin to merge into each other.

A special effort has been made to compress as much information as possible into the illustrations, which have all been redrawn from various original sources in order to clarify and amplify the text, as well as to convey something of that visual aspect of the subject which has a strong appeal for many developmental biologists and which plays a significant part in research. This aspect has been greatly stimulated recently by the rapidly-growing use of scanning electron microscopy as a research technique, and I am grateful to my friends Drs Oliver Flint, Ruth Bellairs and Peter Bryant for allowing me to use original SEM pictures. Permission to use other photographic illustrations has come from Drs G. Gerisch and C. Stolinski, and Mr M. B. H. Mohammed, to whom I am most grateful.

I wish also to thank Mrs M. McCulloch and Mrs C. Morrison for their invaluable help in preparing the manuscript and illustrations. But my chief debt is to all those developmental biologists whose research I have discussed, with many of whom it has been my great good fortune to work or correspond—the latter often as a result of receiving and editing their original manuscripts during a fairly long period as editor of *The Journal of Embryology and Experimental Morphology*. Finally I should like to pay tribute to the happy memory of C. H. Waddington, whose influence will be evident throughout this book.

DONALD A. EDE

Contents

Chapter 1. **LEVELS OF COMPLEXITY IN DEVELOPMENT**   1
Introduction. The genetic programme. *Acetabularia*: development in a single cell. Development in multicellular organisms. Complexities in insect development. Development in cellular slime moulds.

Chapter 2. **FORMATION OF GAMETES AND INITIATION OF DEVELOPMENT**   15
Origin and migration of the germ cells. Spermatogenesis. Role of the sperm in activating the egg. Oogenesis. Amassing genetic information in the oocyte. Seasonal dimorphism in oogenesis. Parthenogenesis.

Chapter 3. **BEGINNINGS OF DEVELOPMENT**   27
Consequences of fertilization. Types of cleavage pattern. Equivalence of somatic cell nuclei. Diversity of the egg cytoplasm.

Chapter 4. **INTERACTIONS IN EARLY DEVELOPMENT**   39
Cellular interactions in echinoderm development. Mosaic and regulative aspects of development. Cytoplasmic factors and gradients in insect embryos. First entry of paternal genes into development.

Chapter 5. **CELLULAR ACTIVITY IN THE EMBRYO AND *IN VITRO***   48
Development as socially-organized cellular activity. Re-aggregation of dissociated cells in culture and in development. Cell movement in embryos. Cell movement *in vitro*. How far does a cell's behaviour *in vitro* reflect its activity in the embryo?

Chapter 6. MORPHOGENETIC MOVEMENTS IN
EARLY EMBRYOGENESIS   62
Cell rearrangements and presumptive-fate maps. Gastrulation. Gastrulation in echinoderm embryos. Gastrulation in amphibians. Early morphogenetic movements in the avian embryo. Morphogenetic movements in insect embryos.

Chapter 7. NEURULATION AND THE DEVELOPMENT
OF THE EMBRYONIC AXIS IN
VERTEBRATES   77
Neurulation in amphibia. Gastrulation, neurulation and axial organization in birds. Cell contacts and cell communication in early chick embryos. Avian neurulation and somitogenesis. Cells of the neural crest.

Chapter 8. DETERMINATION AND
DIFFERENTIATION   92
The molecular basis of cell differentiation. The epigenetic landscape: a model of determination and differentiation. How stable is the determined state? Metaplasia and transdifferentiation. A relation between cell division and determination? Erythropoiesis.

Chapter 9. INDUCTIVE INTERACTIONS   103
Embryonic induction: evolution of a concept. Induction of the vertebrate lens. Instructive and permissive interactions. Induction of metanephric kidney tubules in vertebrates. Primary embryonic induction. The nature of the inductive signal.

Chapter 10. DEVELOPMENT OF THE SKIN AND ITS
APPENDAGES   116
Ectodermal and mesodermal components of the skin. Keratocyte differentiation in mammals. Control of epidermal proliferation. Epidermal-mesodermal interactions in skin development. Cutaneous appendages: hairs, feathers and scales. The development of patterns of feather arrangement.

Chapter 11. MORPHOGENESIS OF A COMPLEX
ORGAN: THE VERTEBRATE LIMB   128
The amphibian limb disc as a morphogenetic field. Determination of polarity in the amphibian limb. Morphogenesis of the avian limb. Cellular activities involved in limb morphogenesis. Ectodermal-mesodermal interactions. Pattern determination along the proximodistal axis. Chondrogenesis in the developing limb bud.

| Chapter 12. | FORM AND PATTERN | 141 |

From the genetic programme to patterns of determination. An approach through automata theory. Developmental signals and cell interactions. Fields, gradients and positional information. Polar co-ordinates in limb regeneration. Alternative theories of pattern determination. Periodic patterns in development.

| Chapter 13. | GENES AND DEVELOPMENT | 155 |

Development and evolution. Genes in individual development. Pleitropy. The use of mutants as experimental tools. The T-locus in the mouse. The *talpid*[3] mutant of the fowl. Gene control systems.

| Chapter 14. | HORMONAL CONTROL OF DEVELOPMENTAL PROCESSES | 166 |

Mammary gland development in the mouse. Phenotypic sexual characteristics. The molecular basis of hormone action. Control of metamorphosis in insects. The epidermal cell in insects. Evidence of gene activation in insect metamorphosis.

| Chapter 15. | INSECT DEVELOPMENT | 176 |

Discontinuities in insect development. Genetic techniques. Cuticular patterns in hemimetabolous insects. Clonal analysis of segmentation in *Oncopeltus*. Imaginal discs in holometabolous insects. Transdetermination in *Drosophila* imaginal discs. Regulation in imaginal discs. Compartments and homeotic genes in *Drosophila* development.

| Chapter 16. | MAMMALIAN DEVELOPMENT | 196 |

Special features in amniote embryos: extraembryonic membranes. Embryogenesis in the mouse: preimplantation stages. Postimplantation stages. Genes and development in the mouse: mosaics. Mouse chimaeras. Coat pigment patterns in chimaeric and mosaic mice.

| Chapter 17. | DEVELOPMENTAL NEUROBIOLOGY | 208 |

Derivatives of the neural tube. Derivatives of the neural crest. Morphogenesis of the neural tube. Relation of the development of motor neurones in the spinal cord to the muscles they innervate. Growth of nerve axons. Orientation of the growing axon. Morphogenesis of the brain. Genetic analysis of neuronal and glial cell development in the cerebellum. Retino-tectal connections and neuronal specificity. Neuronal interactions and systems matching.

| FURTHER READING | 225 |

| INDEX | 242 |

CHAPTER ONE

LEVELS OF COMPLEXITY IN DEVELOPMENT

Introduction

If the egg consists of homogeneous matter as is presumed in this hypothesis (epigenesis), it can only develop into a foetus by a miracle, which would surpass every other phenomenon in the world.

C. Perrault, *Essayes de Physicque*, 1680.

DEVELOPMENTAL BIOLOGY IS THE SCIENCE OF THOSE PROCESSES WHICH result in the increasing structural complexity of organisms in the course of their individual life-histories, or the restoration, in would-healing or in regeneration, of complex structures which have been damaged or lost. Usually, these processes produce the distinctive structural forms which are normal for the species, though very rarely with absolutely precise replication. But sometimes disturbances occur which lead to congenital abnormalities in the embryo, or the development of neoplasms or cancers in late stages, and the study of these fields (teratology and oncology) make developmental biology a science of very great medical importance.

In higher vertebrates, most developmental processes occur in the embryo, before hatching or birth, but in amphibians and many invertebrates changes which take place in postembryonic stages are equally important. These always relate to a change from one mode of life to a completely different one, with a corresponding change in the organism's structure; this process is known as *metamorphosis*, and the transformation from larva through pupa to adult in butterflies and moths and other insects has provided some of the most rewarding material for research and speculation on developmental problems from the beginning of the science in the seventeenth century.

The great attraction and challenge of the subject is and has always been the emergence of complexity from apparent simplicity, of heterogeneity from apparent homogeneity. It was William Harvey, discoverer of the circulation of the blood and in a sense the first modern biologist, who proposed in 1651 that all living things began as "vital primordia" or "ova", and who introduced the term *epigenesis* for his theory that the embryo is not moulded by any external forces but that it is produced by the "motions and efficacy of an internal principle." He admitted that the organism must exist "in potentia" in the ovum, but he concluded frankly that the nature of this potential was quite mysterious, leaving the subject with no explanatory theory for other scientists to follow. Into this vacuum sprang the preformationist theory that the organism existed, with all its parts, in the egg (too small to be visible and requiring only expansion to make it appear) and the complementary theory of "emboîtement" that all succeeding generations were included inside it, one in the other like Russian dolls, and that when they were all used up the species would become extinct. This, to us absurd, theory persisted in one form or another for 150 years, maintained on the most inadequate evidence by naturalists of genius like Bonnet, Trembley and Spallanzani. Though ultimately shown to be hopelessly wrong, it provided a model, as we should now call it, to test against reality and the motivation which led to immense strides in the observational basis of developmental studies. Some of the models described in this book may prove to be equally mistaken, though if so they will not last so long; but their justification will be the same, that they have energized a mass of observations and experiments providing information upon which more satisfactory models may be based.

The genetic programme

What was lacking was an adequate theory of inheritance and a knowledge of its material basis. The birth of modern developmental biology came in the early years of this century with the work of those who realized the connection between their work on the cell, the nucleus, chromosomes and genes and the problems of embryology, especially the American biologists E. B. Wilson and T. H. Morgan, whose books *The Cell in Development and Heredity* and *Embryology and Genetics* appeared in 1925 and 1934. Thereafter the paths of genetics and embryology diverged through concentration on particular aspects of each, especially through embryologists following up the fascinating discoveries of the German biologist, H. Spemann, on the "organizer," which we shall take up later. They were reunited by the work of C. H. Waddington, published in a series of books beginning with *Organizers and Genes* (1940) and it was he who introduced the word *epigenetics* to denote the subject, combining *epigenesis*, the word Harvey used for emergence of complexity from a single primordium, with *genetics*, the new science of inheritance.

With the discovery of the double-helical structure of DNA by Watson and Crick in 1953 and its significance as a genetic code capable of providing a programme for development within the nucleus, this approach became the accepted one, and molecular biologists have established a model for the genetic control of cellular structure and activity which, though its detailed aspects require perpetual updating in the light of ongoing research, is now generally accepted as the basis for controlling the more-complicated processes occurring in the organism during its development. The characteristics of the cell depend upon what proteins (structural or enzymes) are synthesized, and protein synthesis is initiated by construction on the chromosomal DNA of messenger RNA (mRNA: this process is called *transcription*) which codes for a particular protein and which then leaves the nucleus for the cytoplasm, there to control the production of a specific protein molecule *(translation)*.

Acetabularia: development in a single cell

One of the clearest examples of this control has been shown in *Acetabularia*, a marine alga with a stem about three centimetres long with an anchor system of rhizoids at one end and a reproductive cap at the other, whose detailed structure differs in different species, but consisting of only one enormous cell, with a nucleus in the basal rhizoid region. Development is from zoospores produced in cysts which are formed within the cap. If the stalk is cut, the organism is able to regenerate from the base, and grafts can be made between two organisms so that the nucleus-containing base of one can be joined to the stalk and cap region of another, so that even in 1932 Hämmerling was able to perform what were effectively nuclear-transplantation experiments. Anucleate stalks may live for several months, and if the nucleus is removed at an early stage of development, a normal cap can be formed after many weeks.

If the stem, without cap or rhizoids, for one species *(A. mediterranea)* is grafted on to the rhizoid region, containing the nucleus, of another *(A. crenulata)*, a cap is formed which is characteristic of the stem species; but if this is removed, a new cap is formed which is usually intermediate in character between the two species; and if this is removed, a third cap will grow which is characteristic of *A. crenulata* which supplied the nucleus. Thus it appears that factors determining the structure of the cap are present in the cytoplasm of the stalk and survive there over long periods, but that the ultimate source of these factors is the nucleus. There is much evidence

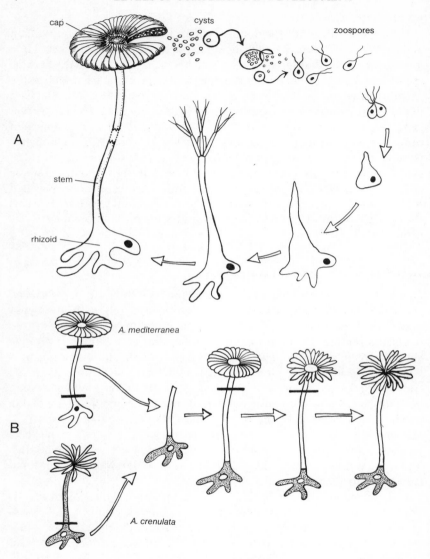

Figure 1.1 Development of the unicellular alga *Acetabularia*.
A. Life cycle of *Acetabularia mediterranea*.
B. Experimental grafting of the stem of *A. mediterranea* to the nucleated rhizoid region of *A crenulata*.
  Based upon J. Hämmerling.

for interpreting this in molecular terms as meaning that control of cap structure depends upon transcription of nuclear DNA into stable long-lived cytoplasmic mRNA. This functions as a signal from nucleus to cytoplasm. However, we have seen that the caps may not be formed for many weeks, even though the mRNA molecules are all present in the cytoplasm, so the timing of cap formation must be controlled by other factors—signals within the cytoplasm—acting at the level of translation.

Development in multicellular organisms

Acetabularia, though we are far from understanding it completely, represents the most elementary type of development, in which rather simple structural differences are produced within a unicellular organism controlled by a single nucleus. In even the simplest multicellular organism, cellular differentiation occurs, i.e. different types of cells are produced with synthesis of different proteins under the control of each cell's own nucleus. As we shall see later, in almost all cases, nuclei in the different cells of multicellular organisms are equivalent with respect to their DNA content, so that it follows that in any cell only a part of the genome is active, and the rest must be repressed. There must therefore be a further intercellular signalling system from the cytoplasm to the nucleus. Furthermore, in order to produce an integrated embryo, the differentiation of one cell must be related to the position it occupies in the embryo in general, and to its neighbouring cells in particular.

Much of current research is concerned with investigating the molecular mechanisms controlling differential gene activity. One approach to developmental biology is the reductionist one that all problems in this field are essentially soluble in these terms, and to concentrate upon simple systems in which analysis at this level is possible; and this approach is particularly appropriate to problems of cell differentiation. But there are also many important problems presented by developmental systems of such complexity that explanations in these terms are unlikely to come within the foreseeable future, where explanations are to be looked for first at other levels of biological activity, e.g. of cellular activity and interaction. These are problems of morphogenesis, of the changing forms of organs and organisms in the course of their development.

Complexities in insect development

This type of problem is well illustrated in the development of holometabolous (higher) insects, such as butterflies, flies and ants. The adults are of immense structural complexity, each part of the body having a characteristic form—from the largest, such as the wings and the legs, down to the minute scales and bristles upon their surfaces. Each of the larger parts consists of complex arrangements of differentiated cells forming an integrated functioning whole. The compound eye, for example, may consist of hundreds of units (ommatidia), each one made up of a complex arrangement of specialized cells and their products—lens, crystalline cone, retinular cells, pigment cells—related to each other and to cells of the nervous system to serve the function of vision. These insects undergo metamorphosis, so that from the egg a sequence of larval, pupal and adult forms is produced. But the rudiment of the compound eye of the adult is present even before hatching, within the embryo, and continues to develop slowly within each larval stage as a minute flattened sac of cells, known as the *embryonic disc*, to undergo a rapid culmination of growth and development and be extruded to the surface along with all the other adult structures in the pupal stage. The orderly way in which the various retinal elements are assembled at the cellular and subcellular level was analyzed by Waddington and Perry in 1960, but in 1970 Kuroda showed that the cells comprising the preommatidial groups could be disaggregated and then reassociated at random in tissue culture, yet still give the beginnings of normal eye development. More recent work on this system which has important implications for fundamental problems in development has been reviewed by Shelton in 1976. A further stage of developmental complexity arises in the many insects, notably termites and ants, living in elaborately organized colonies in which labour is divided among individuals which, though they arise from eggs laid by the same female, the queen, may show enormous differences in size and structure according to the various roles they are destined to fill in the community. This phenomenon, known as *polymorphism* (fascinatingly reviewed and analyzed by Wilson in 1971), presents in a particularly striking form problems of general importance connected with switching from one programme to another in the course of development.

    To discover the genetic programmes underlying events as complex as these requires a type of analysis which goes beyond investigation of mechanisms of differentiation in any particular cell type in much the same way as the analysis of all the products of human activity—social forms, life in cities, the building of the cathedrals in the twelfth century or *Concorde* in the twentieth—must go beyond explanation in terms of individual human lives. Both require analysis at a multiplicity of levels, and of interactions between one individual unit and another, and between groups of units,

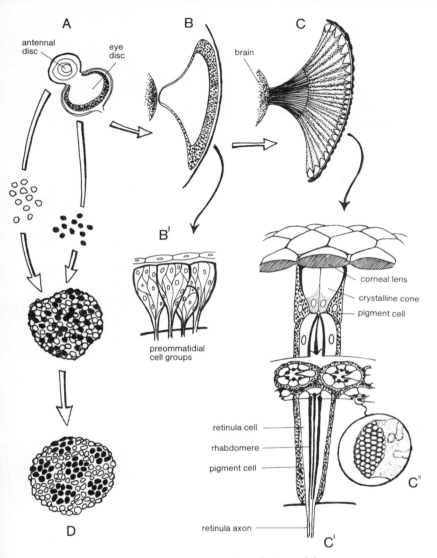

Figure 1.2 Development of the insect compound eye in *Drosophila*.
A. Eye/antennal imaginal disc in the larva.
B. Everted eye disc in the pupa; B^1 preommatidial groups of retinal cells.
C. The adult eye; C^1 A single ommatidium; C^{11} Ultrastructure of a single rhabdomere, produced by self-assembly of hexagonal units originating from infoldings of the cell membrane. Based upon C. H. Waddington and M. M. Perry.
D. Experimental disaggregation and random reassociation of eye (black) and antennal (white) imaginal disc cells, with re-establishment of preommatidial groups in the cultured reaggregate.
Based upon Y. Kuroda.

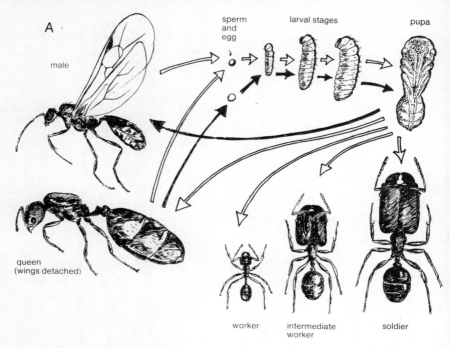
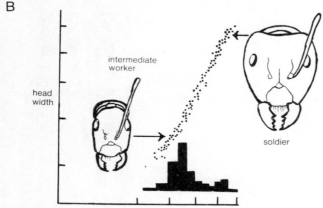

Figure 1.3 Metamorphosis and the development of polymorphic forms in ants.
A. Development of female castes—queens, workers and soldiers—from fertilized eggs, and males from unfertilized eggs, in the Texan harvesting ant *Pheidole*. Based upon W. M. Wheeler.
B. Production of polymorphism through differences in growth and form of the head relative to body size in workers of *Camponotus castaneus*. The distribution curve is bimodal, with a major peak showing that most individual ants are intermediate workers, but there is a second peak at the soldier end. Based upon E. O. Wilson (see chapter 15).

before the behaviour of the whole is explained. The units in the case of developmental biology are almost always individual cells, whose existence, not simply as building blocks but as semi-autonomous individuals with a capacity to exhibit a considerable degree of independent activity when released from the normal constraints within embryonic tissues, is demonstrated clearly in *in vitro* cell culture. The activities which they display there—proliferation by cell division, movement, changes of size and shape, synthesis of one cell product or another—are those which when organized and controlled in the conditions existing when they are in their normal relations to other cells in the embryonic situation produced the sort of complex developmental systems described above. Analysis of this sort then becomes an approach through *cell sociology* rather than simply through *cell differentiation*.

Development in cellular slime moulds

Work on these primitive organisms illustrates the two approaches in a particularly clear way, since their life cycle is divided into a multicellular phase, producing a fruiting body in which cell differentiation occurs in one of only two different ways, as stalk cells or spore cells, and a phase in which the cells exist as independent free-living amoebae, which under conditions of starvation come together in a pattern of social activity which produces the multicellular "slugs" from which the fruiting bodies arise.

In the life-cycle of *Dictyostelium discoideum*, in the amoeboid stage the cells move randomly, feeding upon bacteria, and dividing every few hours. When no food is available, an aggregation phase begins after about eight hours of starvation, and certain cells in the cultures become centres upon which the other cells begin to move in, forming streams which often unite like tributaries of a river, until the cells around each centre have formed a hillock of from 50 to several thousands of cells, depending on the cell density in the culture. The hillock then changes in shape, elongating upwards to form a finger-shaped structure which topples over and forms a slug-like "grex". The grex then begins to migrate, moving in the direction of any light, with its anterior end lifted slightly above the level of the substrate. Finally the culmination stage is reached, when the grex ceases to migrate and produces a vertically growing stalk, with a spherical fruiting body at its tip, within which are produced spores which are eventually liberated and from which new amoebae emerge.

In the transformation from grex to fruiting body, the grex rounds up and

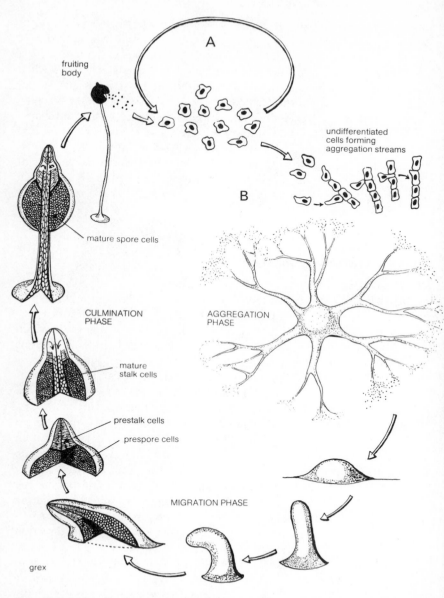

Figure 1.4 Life cycle of *Dictyostelium discoideum* in conditions of (A) ample nutrient supply and (B) starvation.
Based upon J. T. Bonner.

its anterior end becomes a vertical tip from which the stalk develops. As the stalk elongates there is a movement of cells up the outside of the whole structure, and down through the interior to add to the stalk cells, which become vacuolated and encased in cellulose walls; and finally the cells from the rear two-thirds of the grex are left outside at the free end of the stalk to form the spherical mass of spore-forming cells. The grex, then, is divided into an anterior third of potential stalk cells and a posterior two-thirds of potential spore cells. The question arises, when do these two types of cells become *determined* as prestalk or prespore cells, i.e. when do they become committed to these particular fates. There is some evidence that the two types of cell can be distinguished, e.g. by fluorescent immunological staining, before aggregation. If this is so, there must be some sorting out by cell movement within the grex after it is formed, and there is some evidence for this. Certainly the two types of cells can be distinguished by staining techniques within the grex itself, but this commitment is only provisional, for, if the grex is divided transversely into two, each half goes on to produce a fruiting body, complete with both stalk and spores. It appears that the final commitment to one or other paths of differentiation, determination, occurs in relation to the spatial position of the cells within the grex. This means that in addition to the intracellular signals mentioned above, there must be extracellular signals received from neighbouring cells or from the immediate environment.

The simplicity of its system of cell differentiation and the possibility of obtaining genetic mutations (it is a haploid eukaryote, with seven chromosomes) makes *Dictyostelium* excellent material for studies on the molecular mechanisms of differentiation, e.g. in the genetic control of enzymes involved in synthesis of polysaccharides not present in the amoebae but produced in the fruiting body, where controls have been demonstrated at the level of transcription from DNA to mRNA and of translation from mRNA to protein. But the unique advantages offered by the slime mould system are in the opportunities presented for investigations into control of the cell sociology of morphogenesis exhibited so strikingly in the aggregation phase.

It is sometimes necessary to postulate the existence in development of a morphogen, i.e. a signal, consisting of some chemical produced by particular cells in minute quantities, which exerts some controlling effect upon the pattern of morphogenesis, but it has proved possible to extract and define such a morphogen in only one case. This exception is the substance, called on its first discovery acrasin and later found to consist of cyclic adenosine monophosphate (cAMP), which is secreted by those cells

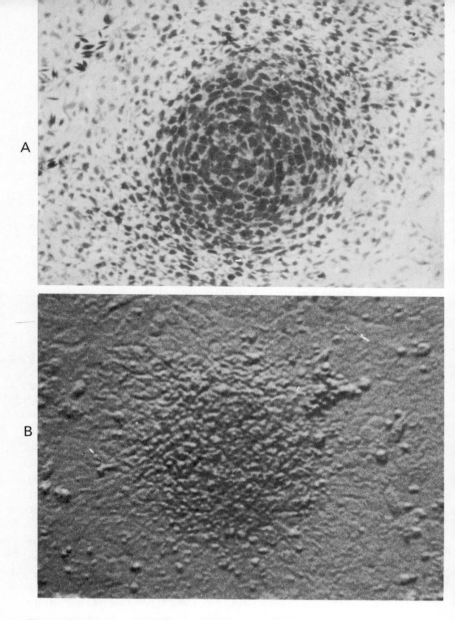

Figure 1.5 Morphogenetic condensations and aggregations in vertebrate and slime mould development.
   A. Transverse section of a digital cartilage condensation in limb mesenchyme of the chick embryo.
   B. Optical shadow-cast photograph of living chick limb mesenchyme cells in *in vitro* culture, aggregating to form a cartilage nodule.

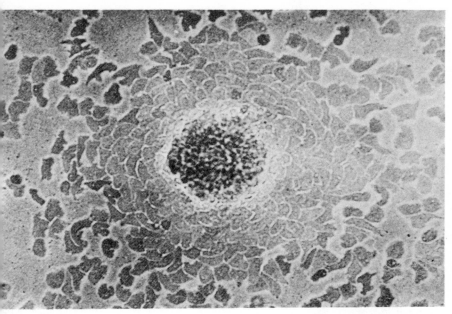

C. Aggregation in the slime mould *Dictyostelium minutum*. (G. Gerisch, in *Current topics of Developmental Biology* (Moscona, A. A. and Monroy, A., Eds.) p. 161. Courtesy of G. Gerisch and Academic Press)

which become aggregation centres and acts as a chemoattractant on neighbouring cells.

Cells entering the aggregative phase begin to secrete cAMP in pulses, i.e. at regular timed intervals, at first independently and at slightly different intervals. Those with the highest frequency become the aggregation centres and become pacemakers for cells around them in an aggregation field, which may be several centimetres across and contain up to 10^5 cells. Secretion of cAMP from the aggregation centre acts as a signal for neighbouring cells, which respond in two ways, by secreting a similar pulse of cAMP and thus relaying the signal, and by making a small movement towards the signalling cell. There follows a refractory period in which the cell cannot respond in either of these two ways, followed by recovery of sensitivity, so that the result is a pulsed movement of cells in the field towards the centre. This movement can be shown in films made using time-lapse cinéphotography, where the cells can be seen in concentric rings moving centripetally in regular bursts, or more often, because of slight perturbations in the wave propagation, in spirals around the centre.

In *Dictyostelium discoideum* the process is complicated by the nature of the contacts made by the aggregating cells, which leads to the formation of the streams and tributaries mentioned above. The preaggregation amoebae are not very adhesive to each other and move by producing cytoplasmic extensions (pseudopodia) randomly around the cell boundary. When they enter the aggregative phase, the cells become polarized, i.e. elongated and moving by production of pseudopodia at one leading edge only, and become attached to each other by strongly adhesive contact sites which occur at the front and rear of the cell. This produces a follow-my-leader behaviour which leads to the formation of the streams, and any cell which comes into a stream from the side has to nose in by waiting for the cell it has come in contact with to move forward until the "tail" is presented, when it can squeeze into the stream.

The slime moulds show very clearly how a stage in morphogenesis can be brought about by the organization of activity in a cell community through interaction by signalling between individual cells. Events of this sort are almost certainly involved in many morphogenetic mechanisms in more complicated organisms, but no system has yet been discovered in which they can be demonstrated. The closest process in appearance to aggregate formation in slime moulds is the formation of cartilage nodules from vertebrate precartilage cells *in vitro*, where clusters of concentrically arranged cells, expanding centrifugally, arise within a uniform cell culture. These clusters closely resemble the aggregations formed by another slime mould, *D. minutum*, in which there is chemotaxis through a cAMP relay system, but no complication by contact following behaviour, so that cells move in on the centre uniformly from all directions and not in streams. The appearance of histological sections through the precartilage condensations of cells in the limb which later produce the skeletal elements is similar, suggesting that there may be a comparable form of aggregation and organization within the embryo, though the visual analogy may be misleading. It may be said, however, that wherever a distinct pattern of cell rearrangement precedes a morphogenetic change in an embryo, some system of cell-cell signalling is usually involved, and the investigation of such systems and the organizational principles underlying them is one of the most important aspects of research in developmental biology.

CHAPTER TWO

FORMATION OF GAMETES AND INITIATION OF DEVELOPMENT

Origin and migration of the germ cells

IN THE MULTICELLULAR ANIMALS, DEVELOPMENT ALMOST ALWAYS STARTS from a single cell, the egg, in which the process is set in train through the act of fertilization by a sperm. Both types of gamete—eggs and sperm—originate as descendants of the primordial germ cells, which are set aside from all the other cells of the body, the somatic cells, very early in the development of the organism.

This early segregation of cells which are going to give rise to germ cells comes about (it is known in many cases in both invertebrates and vertebrates) through the occurrence of special regions in the cytoplasm of the egg (germ plasms) which cause the cells which come to include some of this cytoplasm to differentiate as germ cells or, in some cases, accessory cells connected with them. For example, in insects the egg is torpedo-shaped, with the fertilized egg nucleus situated towards the anterior end; its first divisions produce an expanding sphere of nuclei, each of which moves outwards towards the cortical cytoplasm which invests the nuclei when they arrive just below the surface of the embryo. In higher insects the cortical cytoplasm at the posterior pole, the pole plasm, is characterized by the presence of fine cytoplasmic granules, and those nuclei which arrive in this region (and become invested by pole plasm) later produce the germ cells. If the pole plasm is eliminated by experiment, e.g. by UV irradiation, or through a heredity defect as in the *Drosophila* mutant *grandchildlessness*, the development of the somatic cells will continue, giving a normal adult, but this adult will carry no germ cells and will therefore be sterile. Among vertebrates, the existence of germ plasm at the extreme vegetal pole of the embryo has been demonstrated in the frog *Xenopus* by UV irradiation of this region at the 2 or 4 cell stage, causing sterility but no other defect in the adult frog.

In some insects, the gall wasps, and in some other organisms, such as the nematode *Ascaris*, the chromosomes in all the somatic cell nuclei undergo

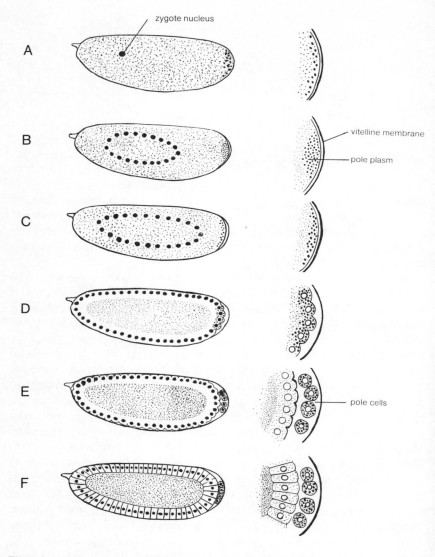

Figure 2.1 Early stages in the embryonic development of the fruit fly *Drosophila*.
A. Egg with zygote nucleus; B-C. Movement of dividing nuclei towards the surface; D. Arrival of nuclei at cortical cytoplasm and passage of some nuclei through the posterior pole plasm; E. Formation of a syncytial blastoderm and a posterior group of pole cells; F. A cellular blastoderm is formed by cytoplasmic cleavage.
Based upon D. A. Ede and S. J. Counce (see chapter 5).

diminution, i.e. fragmentation and loss of fragments, at an early embryonic stage, but chromosomes in the germ cell nuclei do not undergo diminution and must be protected from this process by some factor which is present in the pole plasm but absent elsewhere.

The primordial germ cells are frequently formed a long way away from the gonads (the parts of the embryo in which they settle to complete their development) and they must make their way there by some process of active migration or passive transportation, or both. Thus in the chick embryo the primordial germ cells arise in the posterior margin of the disc of embryonic cells which appears even before incubation, and from there migrate to a position in a crescent-shaped area of extraembryonic cells situated anterior to the head-fold of the early embryo. Then they make their way between the endothelial cells of the extraembryonic blood-vessel walls and are carried with the circulating blood until they reach the neighbourhood of the developing gonads. There they leave the blood vessels and move into the gonads, probably by active amoeboid movement and possibly orientated towards them by chemotaxis.

Once in the gonads the primordial germ cells proliferate, and from their descendants either sperm or eggs are produced. Which pathway is followed depends upon the sex of the embryo, through interaction with the epithelial lining of the gonad, irrespective of the genetic sex of the germ cells. If genetically male germ cells are transplanted to gonads in female embryos, they will produce eggs, and female cells produce sperm. The main stages of development are the same in each—of proliferation, growth, maturation and differentiation—but the details and the cells they produce are very different.

Spermatogenesis

In spermatogenesis the proliferating cells are called *spermatogonia*, which are diploid, having a paternal and maternal set of chromosomes. In many animals (and in all vertebrates) they are found in seminiferous tubules and move in towards the central lumen as they progress through the various stages. Throughout the fertile adult life of the animal, some spermatogonia cease dividing and become primary spermatocytes. These enter the growth phase (though there is in fact very little growth) and following that the maturation stage, in which the number of chromosomes is halved to give the haploid number in the process of meiosis, in which two cell divisions are accompanied by only one cycle of DNA replication, with recombination of maternal and paternal genes. The first cell division of meiosis produces two

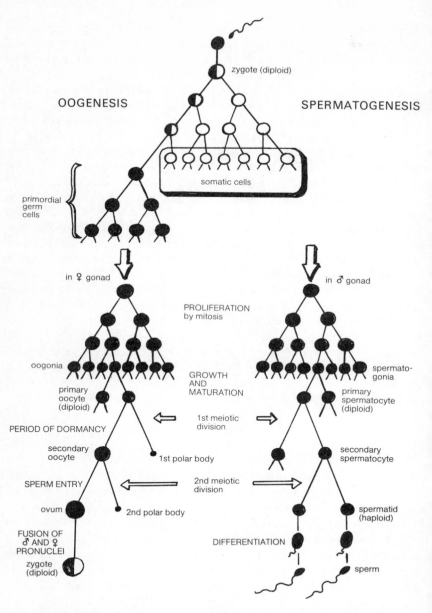

Figure 2.2 Diagrammatic separation of cells into those of the soma and germ line in development, and the production of eggs and sperms from the germ line in gametogenesis.

secondary spermatocytes, and the second division produces two spermatids from each, so that four spermatids arise from each primary spermatocyte. Enormous numbers of sperm are produced, so that in man, for example, each lumen ejaculate contains approximately 15 million. The characteristic features of the mature sperm are acquired in the differentiation phase, when the spermatid is transformed into the highly specialized cell whose functions are to activate the egg to commence its development, and to introduce the paternal genetic material into the egg, in the act of fertilization. The differentiation of the sperm illustrates in itself how complex this aspect of development (the changes which occur within the internal structure of a single cell) may be.

In almost all animals the sperm takes the form of a prominent head and a smaller midpiece, to which is attached a highly active flagellate tail, but in a few invertebrates amoeboid sperm occur. Even within the normal type there is a great variety of form, especially in the shape of the head, which may be nearly spherical, or very elongated, or hooked. The head consists essentially of two structures: the cell nucleus, which undergoes a dramatic reduction in volume, brought about by loss of all its fluid content, so that it is made up almost entirely of the condensed chromosomes, in which little but their DNA is present; and the acrosome anterior to it, a vesicular region of cytoplasm which is specialized for making contact with the surface of the egg at fertilization and undergoing the so-called acrosomal reaction which leads to fusion of the two cells. In the midpiece there are mitochondria which are involved in liberation of the energy required for sperm activity, and two centrioles which are the assembly site for production of the microtubules which support the whip-like sperm tail. The ultimate arrangement of these mictrotubules is as in flagellate structures in general, with two central single tubules at the centre of a ring of nine paired tubules. Differentiation of groups of sperm, up to 74 in mammals, is precisely synchronous because, in the spermatogonial divisions which have produced these groups of cells, separation has not been complete, so that they remain in cytoplasmic continuity and the signals controlling the various steps of differentiation must be common to all cells in the group or (as it may be called since it arises by division from a single progenitor cell) the clone.

Role of the sperm in activating the egg

The means by which the sperm makes its way actively, or is transported passively, to the neighbourhood of the egg surface vary widely in different animals, but the number of sperm which get so far usually represent a minute fraction to those shed by the male. Thus in the rabbit, of the several million sperm deposited in the female vagina, only about 50 reach the egg surface. Of those that do (in most species) only one will penetrate the egg surface, since one of the reactions to sperm entry is a membrane change which creates a so-called block to polyspermy. In rare cases a high proportion of sperm are successful, and most obviously in the honey-bee, where the queen is fertilized only once, during the nuptial flight, yet subsequently produces about $\frac{1}{4}$ million eggs which mostly become fertilized, requiring about 1 in 30 of the original sperm. In insects the egg is enclosed in an impenetrable chorion, through which the sperm can pass at one point only, at the micropyle, whereas in most animals the sperm may make contact with the egg at any point on its surface, though in many it has first to pass through some sort of mucoprotein jelly coat.

There are usually a number of biochemical reactions between the two gametes when the sperm is in the vicinity of the egg, and the most universal of these is the triggering of the acrosome reaction, which leads to actual fusion of the two cytoplasms. It has been described in both vertebrates and invertebrates, and particularly clearly by Colwin and Colwin in the polychaete marine worm *Hydroides*. Here the reaction begins as the sperm comes into contact with the outer covering egg, when the membrane surrounding the acrosomal vesicle and the sperm plasma membrane overlying it break down and fuse at their free edges with each other, resulting in the contents of the acrosomal vesicle now being turned into the surrounding medium, outside of the sperm head. These contain enzymes which break down the egg covering; and projections of the acrosomal vesicle membrane in the region of what was its posterior wall when it was intact, now grow forward, make contact with the egg membrane, and fuse with it so that the egg cytoplasm and sperm cytoplasm are connected. A projection of the egg surface, the fertilization cone, arises to meet and later engulf the sperm, and by further breakdown and subsequent fusion the egg and sperm plasma membranes become continuous as the sperm is drawn into the egg. Fertilization has now been accomplished and is followed rapidly by characteristic changes in the egg which will be described after we have described the development of the egg in oogenesis.

Oogenesis

The sperm is a device for triggering development and for adding paternal genetic material which, as we shall see, can in some circumstances be

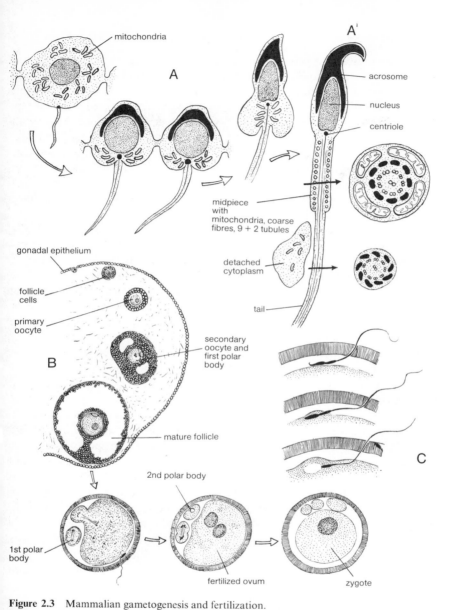

Figure 2.3 Mammalian gametogenesis and fertilization.
A. Development of sperm, showing changes in the spermatids which produce the mature sperm (A^1) in the mouse; B. Development of the egg, showing growth of the oocyte within the ovarian follicle, its discharge from the follicle and maturation with production of polar bodies in the mouse; C. Fertilization in the mouse. Based upon R. Rugh, and C. R. Austin and A. W. H. Braden.

dispensed with altogether. But the egg or ovum must contain all that is necessary for the production of an embryo, including the complete genetic programme for development and molecular machinery connected with it, and also sufficient yolk material to provide for growth and energy requirements up to the completion of embryogenesis, or to a stage when some alternative source is established. This requirement means that the growth phase in oogenesis is one in which a much more dramatic size increase occurs than in spermatogenesis, differing in its extent according to the type of embryo. In vertebrates there is a wide range; in amphibians there is a moderate amount of yolk, producing an egg about 2 mm across; in birds there is a relatively enormous quantity, producing an egg about 1.5 cm in diameter in the fowl; in mammals, where from an early stage the embryo is supplied with nutrients and oxygen through its close, virtually parasitic, association with the mother, there is very little yolk and the egg is small, about 100 μm. Yolk is often distributed non-uniformly, towards the vegetal pole of the egg, while the nucleus is located at the opposite animal pole. These variations in amount and distribution of the yolk lead to major differences in the manner of embryonic development.

Division of the primordial germ cells within the embryonic ovary produces oogonia, which proliferate by mitosis, but which do not in most vertebrates (and in contrast to the spermatogonia) continue to do so after embryogenesis is completed. Far fewer eggs are required than sperm, and all oogonia which are going to become primary oocytes do so, entering the growth phase and the prophase of meiosis which begins the maturation phase by the time of hatching or birth, so that the upper limit of egg production is set at that time. In fact only a small proportion of these primary oocytes continue their development—the remainder die and degenerate in adult life. Those that do continue go through the two cell divisions of meiosis, which produce only one egg. The first division produces the secondary oocyte and a very small cell, the first polar body; in the second division the secondary oocyte divides to form the egg, with the haploid chromosome number, and a second polar body; and in some cases the first polar body divides again, but all polar bodies degenerate later. This second meiotic division takes place after the secondary oocyte is discharged from the ovary, and sometimes only after the sperm has fused with the egg cell.

The passage of yolk and genetic material into the oocyte in the growth phase is usually by way of accessory cells, which may be follicle cells or nurse cells. Follicle cells are derived from the gonadal epithelium and surround the developing oocyte, secreting at a later stage some of the egg membranes, including the fine vitelline membrane which comes into close contact with the cytoplasmic plasma membrane. Materials—lipids and proteins—from which the yolk is synthesized within the growing oocyte pass in through these follicle cells, which are extremely closely connected to the oocyte by interdigitating microvilli, the molecules having been produced elsewhere, e.g. in mammals as far away as the mother's liver. It appears that this is accomplished in many cases by actual ingestion through pinocytosis of these fine cytoplasmic extensions, enabling large molecules to be transferred. It is possible, following observations by Raven on the relationship of follicle cells to the location of specialized subcortical plasms in the egg of the snail *Lymnaea*, that follicle cells may also impose a particular pattern of such plasms on the developing oocyte.

Amassing genetic information in the oocyte

The specification of proteins by production of messenger RNA upon the DNA gene template has been referred to above. Once in the cytoplasm this mRNA becomes associated with ribosomes, made up of ribosomal RNA (rRNA) and protein, and specifies the assemblage there in the correct sequence of amino acids which are conveyed there attached to molecules of transfer RNA (tRNA); the two further types of RNA, like mRNA, are produced on the chromosomal DNA template, but on specific RNA genes. In order to carry development of the embryo through its initial stages, the cytoplasm of the oocyte must be stocked with all three types of RNA. The mRNA, coding for proteins, carries the essential genetic programme and is therefore referred to as informational RNA; its molecules are often temporarily masked by a protein coat and are then known as *informosomes*. Relatively very little of this informational mRNA is required; on the other hand rRNA, which constitutes the main part of the chromosomal assembly machinery, must be present in very large amounts, about 95% of all RNA. Not much tRNA is required—about 4% of all RNA, being only 1% of informational molecules. In the growth plan of oogenesis then, an enormous amount of mRNA must be produced and accumulated in the oocyte cytoplasm, and production in such quantity is beyond the synthetic capacity of the nuclear machinery present in most cells. Special adaptations are therefore found; either in the oocyte nucleus itself or in special accessory cells called *nurse cells*.

Nurse cells are present in the ovaries of many insects, either grouped together with the oocyte or connected with it by long cytoplasmic connections, but in either case in cytoplasmic continuity with it by canals or

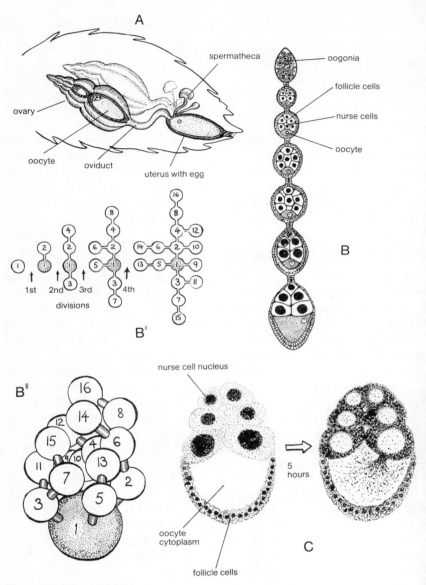

Figure 2.4 Insect oogenesis.
A. Production of eggs in the ovarian tubules of the female *Drosophila*; B. A single tubule showing the production of nurse cells as well as oocytes, illustrated diagrammatically in B[I] and B[II] in which the order of cell divisions is indicated. Based upon K. S. Gill, and E. H. Brown and R. C. King. C. Autoradiographic experiment indicating the passage of ^{3}H-cytidine-labelled RNA from nurse cells into the oocyte. Based upon K. Bier.

bridges. RNA produced in the nurse cell nuclei is transferred through these connections to the oocyte; in the fruit fly *Drosophila* this has been demonstrated by showing that RNA labelled with ^{3}H-cytidine in the nurse cell nuclei is found 5 hours later pouring into the oocyte cytoplasm. The nurse cells are descendants of the same oogonium as the oocyte they are associated with, and the cytoplasmic connections are established, as we have seen in spermatogenesis, by failure to separate completely after mitotic division; in the case of *Drosophila* four divisions produce 15 nurse cells and one oocyte in each group.

In vertebrates there are no nurse cells and this massive RNA synthesis is carried out directly within the oocyte nucleus. The process has been studied most extensively in the frog *Xenopus*. Here, somatic cell nuclei contain two nucleoli, one attached to each chromosome of a particular pair in the diploid set; the nucleolus contains all of the genes coding for rRNA, and synthesis of the latter is therefore restricted to this site. But in the growing oocyte in *Xenopus* there are up to 1200 nucleoli, all of them actively synthesizing rRNA which is discharged into the cytoplasm. Informational RNA is also produced in large quantity, though very much less than rRNA, and this mRNA is synthesized on the rest of the chromosome complement, which is much modified in the oocyte to form the so-called *lampbrush chromosomes*. In these the appearance which led to the name is produced by central regions of the chromosomes being extended as paired loops, each single loop representing a single DNA strand, upon which mRNA is synthesized within the oocyte during the rest of its growth; but most will be stored as inactive informosomes until the early stages of embryonic development.

Seasonal dimorphism in oogenesis

Polymorphism in development has been referred to above, and the origin of the switch into one developmental pathway rather than another may depend upon differences in oogenesis. Thus in the ant *Formica polyctena* the eggs which are produced in winter are much larger than those produced in summer, and this is reflected in the size of the nurse cell nuclei and the amount of RNA they synthesize and pass on to the oocyte. The large winter nurse cell nuclei produce more RNA, manifested most obviously in the form and size of the pole plasm at the posterior end of the egg; the small summer nuclei produce much less RNA and degenerate earlier. By adjusting the temperature, both types of egg can be obtained at the same time, and the amount of food subsequently given to the larva controlled by regulating the number of worker ants allowed to feed them. With a moderate food supply, female larvae from the large eggs develop into queens, and those from small eggs become workers; but this may be partly overriden by subsequent events. If larvae from the winter eggs are fed poorly, or larvae for summer eggs are fed too well, intercastes with a form midway between queen and worker are produced.

Parthenogenesis

The sex of almost all animals is determined according to whether the zygote nucleus contains two X chromosomes or an X and a Y, one parent producing gametes carrying X chromosomes only, and the other producing gemetes of two sorts, half carrying an X and the other carrying a Y; except in birds and butterflies and moths it is the male which is digametic. But in ants, bees and other social hymenoptera, sex is determined according to whether the egg is fertilized by a sperm and is therefore diploid, in which case the embryo will be female, or develops parthenogenetically, without fertilization, to give a haploid embryo which will be male. Parthenogenetic development as a normal event such as this is rare, but artificial activation of the egg in the absence of sperm has long been known, e.g. in the frog in which Bataillon in 1910 produced it by pricking the egg with a needle dipped in blood. In turkeys, selection of very rare spontaneous occurrences has produced genetic lines in which it occurs regularly. In most cases the diploid number of chromosomes is restored early in embryonic development by eliminating a cell division after chromosome replication, and even so development often stops short of completion. But it is clear that the role of sperm in activating the egg is that of a stimulus, for which other quite unspecific physical and chemical agents are as effective.

CHAPTER THREE

BEGINNINGS OF DEVELOPMENT

Consequences of fertilization

ACTIVATION, EITHER BY SPERM ENTRY OR SOME OTHER TRIGGER, IS followed rapidly in those species where observations have been made (chiefly amphibians and sea-urchins) by a number of events which vary somewhat from one organism to another, but which probably occur in some form or other in all activated eggs. Almost immediately, there are two events which may be associated with preventing the entry of more sperm, and which can be observed in the sea urchin: a wave of cortical birefringence observable in a polarizing microscope spreads from the point of sperm entry to cover the whole egg about 20 seconds later, following which a fertilization membrane is elevated from the surface of the egg. The change in birefringence is accompanied by changes in membrane potential and permeability to small ions. Formation of the fertilization membrane depends upon the breakdown of so-called cortical granules, which are in fact ultrastructural vesicles lying immediately beneath the egg plasma membrane; following the passage of the birefrigence wave, the vesicles fuse with the plasma membrane and discharge their contents, forcing the plasma membrane and vitelline membrane apart, and fusing with the latter to form a composite fertilization membrane which is lifted off the surface of the egg by fluid pressure. Proteases are liberated by the breakdown of the cortical granules, and these contribute to the block to polyspermy by destroying sperm binding sites. These changes at the surface of the egg are followed by a general awakening of metabolic activity, especially of protein synthesis, within the cytoplasm.

The nuclei of both sperm and egg (called at this stage *pronuclei*) move actively within the cytoplasm to approach each other; if, as in many species, oocyte maturation is still not completed, the final meiotic division takes place, and the second polar body is extruded; finally male and female nuclear membranes break down and the pronuclei fuse, or the nuclear membranes remain intact until the first embryonic nuclear division begins.

This, with the subsequent cytoplasmic division to give two cells, or *blastomeres* as the early embryonic cells are called, initiates the next stage of development known as *cleavage*.

Types of cleavage pattern

Cleavage results in the partitioning of the cytoplasm of the fertilized egg into blastomeres, each with its nucleus; but at this stage there is no increase in overall mass, so that, while the embryo remains the same size, the cells of which it is made become smaller. A very simple type of cleavage pattern occurs in mammals, where there is very little yolk to distort it, in which repeated cellular divisions produce a solid ball of cells of uniform size, called the *morula*. In most multicellular organisms a space known as the *blastocoele* appears in the interior of the cell mass, and the embryo at this stage is called a *blastula*. Sometimes the blastocoele is central, but in amphibians it is displaced towards the animal pole, where the blastomeres are small, by the much larger yolky blastomeres of the vegetal hemisphere. In amphibians, e.g. in *Xenopus*, the first cytoplasmic cleavage plane is in the plane of the animal-vegetal pole axis, and the second cleavage is in the same plane at right angles to it, giving a four-celled embryo; the third cleavage is in the equatorial plane, displaced slightly by the yolk towards the animal pole, giving an 8-celled embryo. Cleavages follow in a strict order for some time, in the A-V axis plane, then at right angles to it, according to the orientation of the mitotic spindles. The cleavage furrows form most rapidly in the non-yolky animal hemisphere, and divisions in the vegetal hemisphere soon lag behind, giving a blastula with small micromeres at one end and large macromeres at the other.

A characteristic type of cleavage pattern is found in several invertebrate groups, including the snails and other molluscs and the annelid worms, in which the cleavage planes lie obliquely to each other, producing so-called *spiral cleavage*. Usually the first two divisions are at right angles to each other in the plane of the A-V pole axis, and give four equal cells, but the third cleavage is in a plane above the equator, giving four small micromeres rotated by the angle of cleavage on four larger macromeres. Cleavage divisions follow in such a regular sequence, and with such strict angular arrangement, that embryologists at the beginning of the century were able to construct definite cell lineages relating any cell in a late embryo to a particular micromere or macromere, and to give labels to each individual cell defining its place in this relationship. Thus the four first-formed blastomeres are labelled A, B, C and D; the division of A gives a

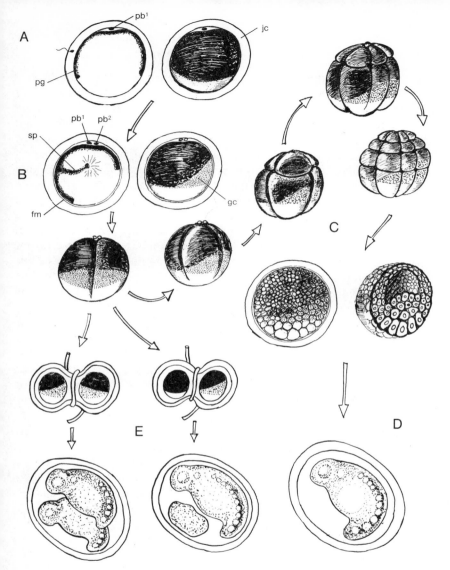

Figure 3.1 Fertilization, cleavage and regulative development in the frog.
A. Sectional and external views of the secondary oocyte at the time of sperm entry into the surrounding jelly coat (jc), showing cortical pigment granules (pg) and 1st polar body (pb¹).
B. Fertilization, showing penetration of the sperm nucleus with sperm path (sp) traced by trail of pigment granules, formation of the fertilization membrane (fm) and the grey crescent (gc), and extrusion of the second polar body (pb²).
C. 1st–5th cleavage divisions, and the blastula in sectional and external views.
D. Fully-developed embryo.
E. Separation of the two first-formed blastomeres, giving regulative development leading to the production of two normal embryos when each blastomere contains grey crescent material.

macromere 1A and a micromere 1a; the macromere 1A divides to give another macromere 2A and another micromere 2a, while the micromere 1a divides to give two micromeres, $1a^1$ and $1a^2$, and so on. Raven has recently shown that it is possible to devise a computer simulation which will reproduce the cleavage of the pond snail *Lymnaea* extremely accurately, programming for order and angle of division. This highly inflexible type of cleavage is called *determinate*, as opposed to the indeterminate type found in the amphibians, in which the fate of individual cells cannot be followed in this way, and in which chance plays a much greater part in partitioning the egg cytoplasm.

Where there is a great deal of yolk, the cleavage furrows may not completely divide the cytoplasm; this is the case in birds, in which cleavage produces a disc of embryonic material at the animal pole. Insects show quite a different type of early embryonic development within a highly yolk-charged egg, which should strictly not be called *cleavage* (though by custom it usually is) since the first nuclear divisions (the first 12 in *Drosophila*) occur in the complete absence of cytoplasmic division. The zygote nucleus lies in the interior of the egg, nearer the anterior end, surrounded by a fine cytoplasmic network whose vacuoles are packed with large yolk granules; at the periphery of the egg the cytoplasm forms a yolk-free cortex. When the zygote nucleus divides, the daughter nuclei, each surrounded by a thin layer of cytoplasm, move apart and divide repeatedly, producing an expanding sphere of nuclei and their surrounding cytoplasmic layers, which eventually reach and enter the cortical cytoplasm. Once arrived there they continue dividing until the cortex is packed with a single layer of nuclei, forming a syncytial blastoderm immediately below the vitelline membrane, except at the posterior region where the pole cells (the germ cells) form a separate group between the blastoderm and the vitelline membrane. After the twelfth nuclear division, cleavage furrows extend inwards from the periphery of the embryo and cut off each nucleus within its own cytoplasm, producing a cellular blastoderm. From now on the divisions, which were synchronous, became asynchronous.

Equivalence of somatic cell nuclei

One of the most notable among nineteenth-century embryologists was August Weismann, who, forced to abandon his microscopical work owing to failing sight, produced a number of influential theoretical studies on development and heredity. One of his suggestions was that differences

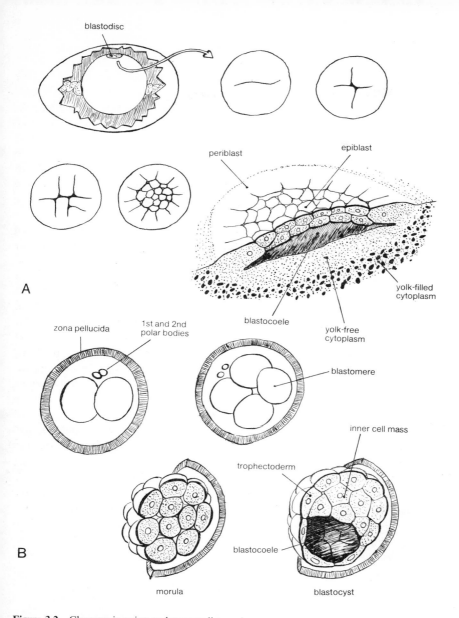

Figure 3.2 Cleavage in avian and mammalian embryos.
A. Cleavage in the blastodisc of the hen's egg, producing a modified blastula—the blastoderm.
B. Cleavage in the mouse. The egg is enclosed in a zona pellucida, within which cleavage produces the morula, which is transformed into a modified blastula—the blastocyst (see chapter 16).

between cells arose through differential distribution of "determinants" on the chromosomes—genes, as we should now say—so that the nuclei in cells of any one tissue type would contain only a special fraction of the chromatin present in the original zygote nucleus. The chromosomal diminution observed in the somatic cells of *Ascaris* and some other organisms provided some support for this hypothesis, but subsequent work has shown that these are special cases which have no bearing on the problem of differentiation between somatic cells, and that it is probably a universal rule that the nuclei of these cells are genetically equivalent and contain all the genes present in the fertilized egg.

The chief evidence for nuclear equivalence is from studies on amphibians, pioneered by Briggs and King in 1952 and in subsequent work on the frog *Rana pipiens*. In 1958 Fischberg, Gurdon and Elsdale developed a highly successful nuclear transplantation technique for the South African clawed toad *Xenopus*, which has subsequently been used by Gurdon and colleagues to explore nucleocytoplasmic relationships in great depth. Nuclei were taken from differentiated tissue cells of a late *Xenopus* embryo or tadpole, or even an adult, and transplanted into eggs whose own nuclei had been removed or destroyed. The clearest results came from transplanting nuclei from intestinal epithelium cells from tadpoles directly and adult skin cells which had been cultured *in vitro* for a few days, using *Xenopus* which was heterozygous for the anucleolate mutant and whose nuclei were therefore clearly marked by having only one nucleolus instead of two. In these experiments the egg nucleus was destroyed by UV irradiation, and the donor nucleus, surrounded by a very thin film of cytoplasm was injected into it; some of these eggs subsequently underwent development up to the tadpole stage in the case of skin cell nuclei, and to completely normal fertile adults from the tadpole intestine nuclei; in all cases the nuclear marker showed that descendants from only the donor nucleus were present. The percentage of successes is small, but can be improved by serial transplantation from blastular cells, which gives development to the tadpole stage from about 20 per cent of intestinal nuclei and 12 per cent of cultured skin nuclei. These results indicate clearly that there is no loss or permanent inactivation of genes in the course of development, and that genes which have no expression in a highly specialized cell within the fully developed organism can be reactivated when transplanted into the egg cytoplasm. The experiments also indicate the immense significance in development of the egg cytoplasm, which is able to reprogramme nuclei from somatic cells from advanced stages of development in such a way that normal development can result. Essentially the same results have since been

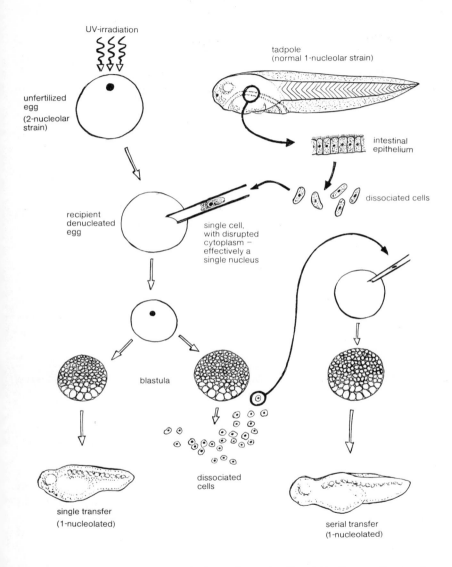

Figure 3.3 Demonstration of the equivalence of somatic cell nuclei in *Xenopus*. Best results are obtained when the first blastula is dissociated and the blastomere nuclei used for further transplantation into enucleated eggs.
Based upon J. B. Gurdon.

obtained in *Drosophila*, using blastoderm cells as the source of donor nuclei.

In the exploration of nucleocytoplasmic relationships, one other technical advance of great potential should be mentioned, though there is not space to do more than draw attention to it here. This is the technique of cell fusion and cell hybridization developed by Harris and Watkins in the 1960s (clearly reviewed and its implications discussed by Harris in 1970). Cells which are cultured in the presence of "Sendai" virus may fuse together, even though they may be highly differentiated—e.g. fibroblasts, macrophages, lymphocytes, nucleated erythrocytes, and (what is more astonishing) even though they may be from quite different vertebrate species, e.g. human, mouse, chick or *Xenopus*, providing material of enormous value in furthering our understanding of nucleocytoplasmic relationships, nuclear equivalence, control of transcription, and other aspects of the molecular events underlying differentiation.

Diversity of the egg cytoplasm

In all animals, the mature oocyte already shows some ordered cytoplasmic diversity, manifested at least as a polarization between animal and vegetal poles. In the common brown seaweed *Fucus*, this is not so; its fertilized eggs are spherical and show no sign of being polarized until they become pear-shaped by the extension from some part of the cell surface of a rhizoidal region which later becomes cut off by formation of the cell wall in the first cleavage. Many factors have been found to affect the direction in which the rhizoidal protuberance grows out, e.g. in a group of eggs in sea-water at normal pH, the protuberances will be towards the centre of the group, but in alkaline sea-water they will point outwards. The stimulus may be very brief, e.g. a flash of light, and this environmental clue must be amplified and stabilized in some way within the egg. Jaffé, using techniques which enabled him to measure the extremely small voltages involved, has produced evidence that this depends upon the production of an electric current through the egg by movement of charged macromolecules to the rhizoidal region.

There is very little evidence as to how polarity is established in animal oocytes, but it must be imposed by influences from the surrounding ovarian cells. In insects, where the egg is usually more or less torpedo-shaped, the main axes of symmetry—anterior-posterior, dorso-ventral and bilateral (right-left)—are established in the oocyte to correspond with those of the maternal ovary within which it is developing; and the elongation of the oocyte appears to be achieved, according to observations by Tucker on oogenesis in cockroaches, by circumferentially arranged microtubles in the follicle cells, joined to desmosomes (a type of cell junction) which acts as intercellular rivets, forming a system which restricts circumferential expansion while allowing it in an antero-posterior direction. Besides the overall shape of the oocyte, there must also be fundamental cytoplasmic differences related to the axes of symmetry.

In amphibians the oocyte is spherical, but the animal-vegetal axis is clearly established by the positioning of the nuclear material and the extrusion of the polar bodies at the animal pole, and concentration of yolk towards the vegetal pole; this polarity is emphasized by the presence of black pigment granules in the egg cortex (the layer of cytoplasm immediately beneath the plasma membrane) in the animal hemisphere, but not in the polar region of the vegetal hemisphere. Establishment of the plane of bilateral symmetry, of the right-left axis, depends upon the position of the point of sperm entry at fertilization. This is clearly marked, since the sperm head traces a path consisting of pigment granules which it carries in with it as it moves towards the egg nucleus. Then follows a rotation of the cortex on the deeper cytoplasm, together with the pigment granules, which leads to the appearance of grey crescent area directly opposite to the point of sperm entry in what was previously a pigmented region of the egg, adjacent to the non-pigmented vegetal region. This grey crescent region becomes the dorsal side of the embryo, and the first cleavage furrow (which divides the embryo into right and left halves) usually passes through its centre.

The grey crescent is a region of cytoplasm possessing special characteristics which, as we shall see later, is essential for normal development in amphibians and is an example of the existence in early embryos of specialized cytoplasmic regions (plasms) related to later development in a particular way. One other such special cytoplasm has already been described—the pole plasm which leads cells within which it is included to be set apart as the primordial germ cells. Plasms may be distinguished by differences of pigmentation, as is the grey crescent; and in *Styela*, an ascidian whose embryology was the subject of a classical study by Conklin in 1905, rearrangement of the egg cytoplasm after fertilization in a manner similar to but more complicated than those seen in amphibians produces a variety of differently pigmented regions whose fate in the developing embryo was traced: a dark yellow plasm contributes to the tail muscles in the larva; a light yellow plasm becomes the coelomic mesoderm; a light grey plasm is incorporated into the notochord and neural plate; and so on.

A particularly interesting example of a highly localized special cytoplasm is the polar lobe found in certain marine molluscs, e.g. the gastropod *Ilyannassa*, investigated by Wilson in 1904 and in great detail by Clement more recently. The polar lobe is a globular extension of the vegetal pole cytoplasm formed first during the first cleavage division, and of about the same size as the blastomeres, so that the two-celled embryo is trefoil-shaped. It is a transient structure, soon withdrawn into its blastomere, but it is formed again at the second cleavage, and so on for several cleavages. Cleavage in these embryos is of the spiral determinate type, and the lobe is always produced from and flows back into the D macromere. Since it is attached by only a very narrow cytoplasmic bridge, it can be removed very easily, and with no damage to the nucleus. Normal development of the *Ilyannassa* embryo leads to the production of a planktonic "veliger" larva but, if the polar lobe produced at the first cleavage division is removed, the veliger develops without any mesodermal structures, suggesting that it is the cytoplasm which periodically forms the polar lobe which contains determinants that lead to expression of genes coding for mesodermal differentiation in nuclei that come to be associated with them. Removing the polar lobe at successively later cleavage stages leads to progressively less interference with normal development, indicating that the determinants present in the original blastomere are distributed into its daughter cells, micromeres 1d, 2d, etc., and their descendants.

We see that in some embryos clear cell-lineages can be observed, and also that determinants localized in various regions of the egg cytoplasm may be distributed in an ordered way through cleavage, causing those cells and their descendants which contain them to undergo particular developmental fates. This clearly provides a possible mechanism for development, providing that a sufficiently precise localization of determinants can be achieved during oocyte development or very early embryogenesis. In development of this type, experimentally isolated blastomeres would develop strictly in accordance with the fate their cell lineage would have undergone within the embryo, and an incomplete embryo would be produced. Wilson found just this to be the case in the marine mollusc *Patella*, and such development is described as *mosaic*, since it consists of the autonomous development of the different regions of an egg or early embryo, fitted together as a mosaic of independent parts. However, in most embryos interactions occur between one region and another such that, if a part of the embryo is isolated, some degree of regulation occurs, leading towards the production in most cases of a complete and normal embryo. Thus, if the two blastomeres produced by the first cleavage division in an

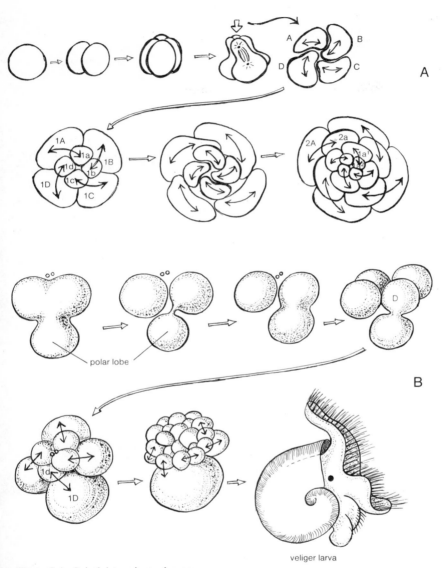

Figure 3.4 Spiral determinate cleavage.
A. Diagrammatically from the side and from the animal pole. Two-headed arrows indicate orientation of mitotic spindles.
B. In the mollusc *Ilyannassa*, showing the appearance of the polar lobe, and the planktonic veliger larval stage which is characteristic of this and other marine snails.
Based upon A. C. Clement.

amphibian egg are separated, both may go on to produce a complete embryo, though containing only half of the original cytoplasm, and with that cytoplasm distributed in a different way. Such development is described as *regulative*, but even here regulation occurs only within limits set by the distribution of certain cytoplasmic determinants. In the amphibian, for example, some grey crescent material must be included in a blastomere for development to proceed normally and if, through abnormal orientation of the first cleavage plane, all of the grey crescent material goes into one blastomere, only that blastomere will produce an embryo. Furthermore, a stage in embryogenesis will be reached when regulation can no longer occur, and removal of a part of the embryo will produce a corresponding defect in its development. The distinction between mosaic and regulative development cannot be clearly drawn—for example, embryos with spiral cleavage are essentially mosaic, yet some interactions between one region and another do occur—and, as we shall see later, analyses of this aspect of development are better conducted in terms of when and how cellular determination takes place.

CHAPTER FOUR

INTERACTIONS IN EARLY DEVELOPMENT

Cellular interactions in echinoderm development

ALL OF THE ECHINODERMS—SEA CUCUMBERS, STARFISH AND SEA URCHINS —have a complex and interesting developmental history in which a freeswimming, planktonic, bilaterally symmetrical larva (called a *pluteus* in sea urchins) arises from the fertilized egg, to be transformed in a subsequent metamorphosis to a bottom-living radially-symmetrical adult with a totally different form, produced within the larval body. Most attention, however, has been paid to the early embryonic stages, and the sea urchin in particular has provided material for important experimental studies. The most historically important were those of Hans Driesch, who showed in 1891 that separation of the first two or four blastomeres led to the formation of small but complete pluteus larvae, demonstrating regulation in embryonic development for the first time, and leading Driesch himself to despair of finding a materialistic explanation and to invoke instead a mysterious guiding force, the *entelechy*. More recent experimental work on the sea urchin has been largely due to Hörstadius, beginning in 1928.

Cleavage leads to the production of a 64-cell stage in which the blastomeres are arranged stacked in circular tiers—two animal tiers (an_1 and an_2)—two vegetal tiers of slightly larger macromeres (veg_1 and veg_2) and at the vegetal pole a tier of eight very small micromeres. Further cell division reduces the size differences and produces a hollow ciliated blastula, then a gastrula, through a process which will be described later, and subsequently the pluteus larva, distinguished by the possession of a gut and a skeleton which supports a system of ciliated arms. Each of the tiers in the early embryo normally produces a particular set of structures in the pluteus. Driesch called these their prospective fates, but his experiments had shown that their prospective potencies were far wider, and that the fate of any blastomere might be determined according to its position in relation to other cells rather than by its cytoplasmic lineage. But Hörstadius showed that the capacity for regulation is limited and that, if the embryo is halved

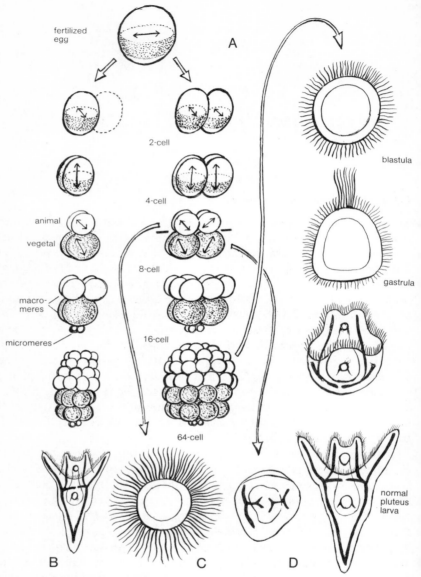

Figure 4.1 Development in the sea urchin.
A. Normal development leading to the planktonic pluteus larva.
B. Development from a half-embryo produced by separating the two first-formed blastomeres, leading to a small but perfect pluteus.
C. and D. Development from half-embryos produced by separating animal and vegetal tiers of blastomeres at the 8-cell stage.
Based upon S. Hörstadius.

horizontally at the 8-cell stage, the animal half develops into a simple blastula-like structure with very long cilia and the vegetal half produces a sac-like structure with no cilia, poorly developed skeleton, and an abnormally large gut with no mouth opening. Yet the result is still different from that obtained in mosaic embryos where we have seen that isolated fragments form exactly what they would have done in the context of the whole embryo, whereas in the sea urchin animal and vegetal characteristics are abnormally exaggerated in isolated halves. And this can be shown, not only by separation at the 8-cell stage, but by dividing the fertilized egg at right angles to the animal-vegetal axis. This suggests that the development of each half is influence by some factor produced by the other half.

By making various combinations of different tiers from a 64-cell stage embryo, it can be shown that the concentration of these factors, presumably chemical substances, varies in a graded way from one end of the embryo to the other, and that the resulting development depends upon the relative concentrations of these factors along the animal-vegetal axis. The results may be explained if it is supposed that a gradient is set up for each factor within the egg cytoplasm, persisting within the early embryo, with the high point of one at the animal pole and its low point at the vegetal pole, and vice versa for the other, and that the gradients are re-established by interaction between blastomeres when they are rearranged, so that the structures of the pluteus larva are produced in the normal order, though not necessarily normal proportions. Thus addition of four micromeres to the isolated an_1 tier will produce a small but otherwise normal pluteus; addition of only two or one gives embryos in which the "animal" structures are progressively more pronounced; similar combinations of micromeres with the an_2, veg, and veg_2 layers give results which this double-gradient model would lead us to expect.

Mosaic and regulative aspects of development

In general then, observations on a wide variety of embryos show two possible mechanisms for the control of early development. In mosaic embryos there are specific determinants localized in the egg cytoplasm, each leading to characteristic cell types, and hence to particular structures. In regulation embryos there is a gradient of some factor, possibly a chemical substance, whose level in different regions of the embryo specifies what cell types and body structures shall be formed; two or more interacting gradients may exist. Highly mosaic development is unusual in

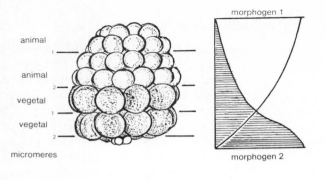
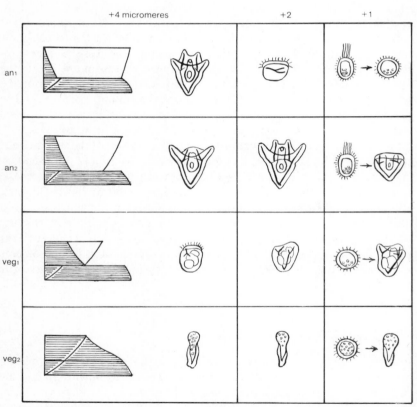

Figure 4.2 The double-gradient model in echinoderms.
Development of various combinations of tiers taken from 64-cell stage sea urchin blastulae and its explanation in terms of interaction between two morphogens whose concentrations vary in opposite directions along the animal-vegetal axis.
Based upon S. Hörstadius.

the animal kingdom, and one would expect that natural selection would generally favour a mode of development which allowed for regulation of any deviations from a rigidly determinate cleavage pattern. Furthermore, it has become impossible to believe that for every type of cell, in every one of its arrangements in the fully developed embryo, there should be a specific and appropriately positioned cytoplasmic determinant in the egg. Neither is it necessary to suppose that the cytoplasm of any animal egg is completely structureless. In all eggs there will be some distinction between one region of the cytoplasm and another, and some interactions between one region of the embryo and another; what further research must look for are the precise control mechanisms operating in particular embryos and especially to investigate their molecular basis.

Cytoplasmic factors and gradients in insect embryos

Some promising progress in this direction has been made by Sander and Kalthoff, working on insect embryogenesis. In insect eggs the yolk is distributed uniformly, so that there is no animal-vegetal axis, but there is a clear antero-posterior axis along which the embryo develops. Even at an early stage the embryo, which forms along the ventral side of the cellular blastoderm, is clearly divided in this axis into easily distinguishable regions which will give rise to the head, jaws, thorax and abdominal segments, so that any changes in their lineal order produced by experiments can be easily seen. It has been shown that the nuclei are equivalent during the multiplication phase in the interior of the egg, so that regional differences in later development must arise through differences established in the cortical layer of cytoplasm which becomes populated by the nuclei and which then by cytoplasmic cleavage gives the cellular blastoderm.

Sander showed that in the leaf hopper *Euscelis*, even more clearly than in the sea urchin, regulation following experimental interference by no means always leads to normal development but may produce curious monsters, notably one in which the anterior structures are replaced by a mirror image of those in the posterior half, giving a "double abdomen" embryo. As in the sea urchin, a double-gradient model explains the facts best, with the highest level of one gradient at the posterior end of the egg and the highest level of the other at the anterior end, with specification of the various parts according to the relative concentration of the two factors along the antero-posterior axis.

The *Euscelis* egg is elongated and may be constricted by a ligature at any point along its length. Furthermore, the posterior tip cytoplasm is marked

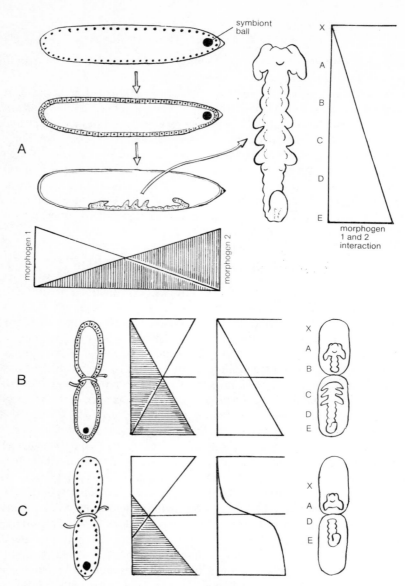

Figure 4.3 The double-gradient model in insects.
A. Normal development in the leaf hopper *Euscelis* showing syncytial blastoderm (late nuclear division stage), cellular blastoderm (late blastoderm) and embryonic germ band stages, with the ball of symbiont bacteria situated posteriorly. The two hypothesized morphogen gradients are indicated below and both are combined in a graph indicating the product of their interaction (right).
B. and C. Development following constriction at the late blastoderm stage (B) and the late nuclear division stage (C).
Based upon K. Sander.

by the presence of a ball of symbiotic bacteria which may be pushed anteriorly into any other position. In Sander's experiments the regions were scored as X (extra-embryonic membrane anterior to the embryo proper), A (head), B (jaws), C (head), D (anterior abdomen) and E (posterior abdomen). Constriction at the late blastoderm stage produces two partial embryos which together make up all the structures of a normal whole embryo, but done at the late nuclear division stage the result is that the parts produce less than they would normally, the middle structures being absent altogether. On the double-gradient hypothesis this follows because the gradients are not yet completely set up, and the interactions producing the central structures do not occur.

If the last experiment is repeated, but the symbiont ball pushed just anterior to the constriction before it is made, the posterior portion will develop defectively as before, but the anterior portion will give a complete though much shortened embryo, since the source of the posterior gradient factor is now present as well as the anterior one. If the constriction is made much further forward, and the symbiont ball pushed up only to its posterior border, the X region is produced in the anterior portion, while in the posterior portion, since the posterior gradient factor is now present anteriorly and still persists posteriorly, a double abdomen embryo is produced with—since the anterior gradient factor is in low concentration here—no interaction sufficient to give the anterior regions. If the symbiont ball is pushed to the middle at the nuclear division stage, and a constriction adjacent to it made only later at the late blastoderm stage, by which time the new gradient is set up through the whole length of the cortex, the result is the same whether the constriction is anterior or posterior to the ball. A complete, though shortened, embryo develops from the anterior position and a reversed half-embryo from the posterior portion.

Kalthoff found that double-abdomen embryos could be produced in the chironomid midge *Smittia* by ultraviolet irradiation of the anterior pole of the egg at the nuclear division and migration stage, as might be predicted on Sander's model if the source of the anterior-determining factor were eliminated. The UV damage, however, may be repaired by subsequent exposure to visible light, which suggested to Kalthoff that the factor was a nucleic acid. This is supported by evidence that injection of the enzyme RNAase at the anterior tip also produces a high proportion of double abdomens in survivors, and it is possible that the factor is in fact maternal messenger RNA, localized in the cytoplasm during oocyte formation, which is translated at this stage.

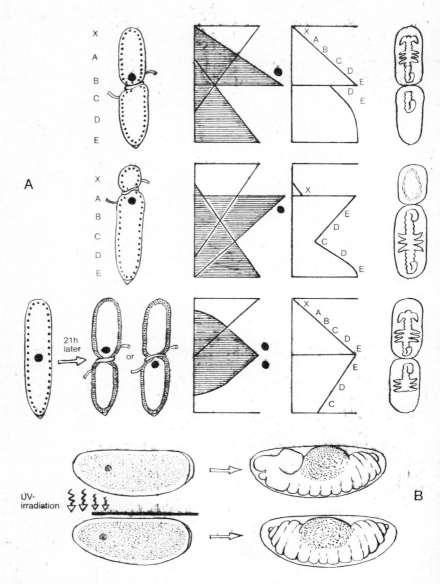

Figure 4.4 Double gradients and double monsters in *Euscelis*.
A. Development following constriction after the cytoplasm around the symbiont ball has been moved anteriorly.
   Based upon K. Sander.
B. Normal development in the chironomid midge *Smittia* and production of a double-abdomen embryo by UV-irradiation of the anterior tip of the egg.
   Based upon K. Kalthoff.

First entry of paternal genes into development

The earliest stages of development, then, are directed by cytoplasmic factors within the egg which have been determined ultimately by the maternal genotype. Many inherited embryonic abnormalities, e.g. in *Drosophila*, will therefore be due to genetically controlled maternal effects upon the egg cytoplasm, of which "grandchildlessness", discussed in chapter 2, is an example. Another is the sex-linked mutant *deep orange* (*dor*) in which embryos from mothers which are homozygous for the *dor* gene develop abnormally and do not survive beyond the gastrulation stage, unless they are fertilized by sperm carrying a *wild-type* (i.e. normal) gene rather than a *dor* gene or simply the genetically inert Y chromosome. Garen and Gehring in 1972 used this mutant to show that embryonic development proceeded much further, to organogenesis, if the eggs were injected with cytoplasm from *wild-type* eggs. Whether the *wild-type* sperm and the *wild-type* egg cytoplasm alleviate the maternal effect for the same reason, which is quite unexplained but might possibly be, for example, through contributing a necessary messenger RNA, such rescue experiments, with the possibility of defining the injected factor, suggest a promising approach in investigating the molecular basis of cytoplasmic control mechanisms.

We have seen that in amphibian oogenesis the oocyte is stocked with informational RNA and the RNA molecules required for its translation, and it can be shown that if all chromosomal material in the fertilized egg is eliminated, development will proceed normally up to but not beyond blastula formation. The same is true for sea urchins, indicating that in these organims no zygote gene control is required until the events of gastrulation. The clearest demonstration of the beginning of zygote gene transcription is the time at which paternal genes produce their effects upon embryogenesis, and in *Drosophila* a number of mutants, described by Poulson in 1945 and by Ede and others subsequently, have been found to exert very much earlier effects, e.g. in disrupting the movement of nuclei into the cortex. Evidently the time at which zygote gene transcription begins varies considerably, but in no organism is it later than gastrulation. The use of mutants which produce clear and often lethal embryonic disturbances provides an important technique for investigating the mode of action of genes in development at all stages, not only in *Drosophila* but in all organisms where genetic material is available.

CHAPTER FIVE

CELLULAR ACTIVITY IN THE EMBRYO AND *IN VITRO*

Development as socially-organized cellular activity

THE DEVELOPMENT OF AN ORGANISM VIEWED AS A WHOLE, ESPECIALLY if it is seen speeded up by time-lapse cinéphotography, has a mysterious quality, rather as if a pot should arise from a wheel in the absence of a potter. It appears as less of a mystery considered as arising from the ordered activity of the cells of which it is composed, just as choreographers of film musicals have created swirling and changing patterns from the movements of hundreds of individual dancers. The problem becomes one of how the activity of each individual cell is so controlled and integrated with that of its neighbours as to produce a harmoniously developing organism. Some developmental processes occur in systems which are not divided into cells, as in the early stages of insect embryogenesis, but these are exceptional and of limited complexity. Generally, division into separate daughter cells follows each nuclear division, and the existence of these cells as semiautonomous individuals (even in adults) is demonstrated in normal conditions in wound-healing, when cells which have appeared static move out to repair damaged tissues—and more dramatically in tissue culture, when cells from small fragments of tissue (explanted on glass or plastic in an appropriate medium) wander out and display behaviour patterns which may be studied in much the same way as protozoan behaviour is studied.

Changes in the overall form of the embryo depend upon many cellular activities: changes in cell numbers arising from cell division; changes in cell size and shape; changes in spacing through production of intercellular products, and in arrangement through movement; even cell death, which is a part of normal development. Of these, movement is the one which is most striking as an indicator of autonomous cell activity and which will be considered now.

Recognition of the degree to which cells do behave as co-operating but independent individuals came first in the sponges, which were often regarded as organized two-layered colonies of essentially separate cells, specialized for various functions, rather than true metazoan organisms. Minchin, in 1898 and subsequently, described the formation of the calcareous or siliceous spicules which give the sponge its supporting structure, sometimes loosely arranged, in other species forming a complex and beautiful skeleton. In *Clathrina*, there are triradiate spicules in the outer layers of cells and quadriradiate cells in the inner layer, with the fourth ray projecting into the gastral cavity. The triradiate spicules are formed by cells which migrate into their correct position and arrange themselves in groups of three, clover-leaf fashion. Each cell then divides, producing two superimposed trefoils, at the centre of which the spicule begins to appear, consisting at first of three separate rays. The rays enlarge and fuse basally to give the triradiate structure, and the inner trefoil formative cells move out along the rays as they grow, leaving them covered by a thin layer of cytoplasm and secreting spicule material as they go. The outer cells build up the base, and when that is completed they too move out to the apex of the spicules. Formation of the quadriradiate spicules begins in exactly the same way but, when the rays have fused and begin to grow out, the group of formative cells is joined by another cell of quite different origin within the sponge, which settles at the base and divides into two cells which secrete and add their ray to the other three, orientated correctly.

Reaggregation of dissociated cells in culture and in development

The recognition of this ability of sponge cells to move about and co-operate with each other led H. V. Wilson in 1907 to look at the effect of dissociating cells of *Microciona* by forcing sponges through fine bolting silk. He found that the cell aggregated together, forming small clumps within which the typical sponge structure was "reconstituted" either, as he thought, by transformation of cells into appropriate types or, more likely as J. S. Huxley proposed when he worked on this problem, by rearrangement of cells by active movement into their appropriate positions. Astonishing as this seemed at the time, it was regarded as a peculiar property of an organism low down in the evolutionary scale. But many years later, largely as a result of work by Holtfreter on dissociated cells and tissues in amphibian embryos, its great general significance was realized. That embryonic cells of higher animals show an equal capacity for reconstitution was demonstrated by Moscona in 1952, when he dissociated cells from a variety of tissues in chick and mouse embryos, using the enzyme trypsin, and recombined them by placing them in tissue culture medium in a gyratory shaker. The cells, brought into close contact through gentle

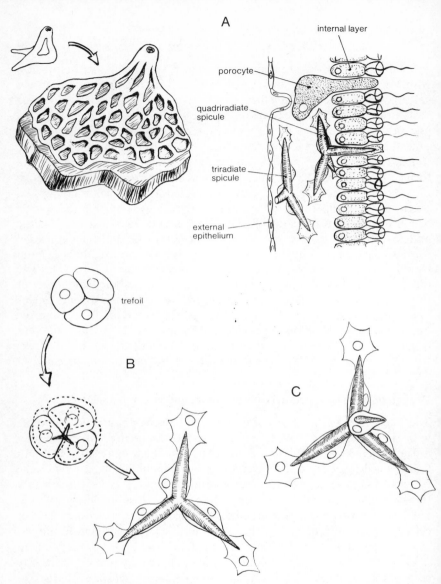

Figure 5.1 Cellular activity in the formation of spicules in the sponge *Clathrina*.
A. Growth of the sponge and a section through part of it, showing triradiate spicules in the external and quadriradiate spicules in the internal layer.
B. Formation of a triradiate spicule from a trefoil made up of 3 cells which have migrated in from the external epithelium.
C. Formation of a quadriradiate spicule by the additional activity of another cell, derived from a porocyte.
Based upon E. A. Minchin.

centripetal force, adhered together to form reaggregates within which their characteristic tissue architecture was organized: chondroblasts which give rise to cartilage in the embryo did so in the reaggregates; kidney-forming cells formed nephric tubules; retinal cells formed characteristic rosettes; salivary-gland cells produced a system of branched secretory ducts. Moscona made recombinations of cells from different tissues and showed that the cells sorted themselves out according to tissue type but, when cells of the same tissue but from different species were mixed, they produced a chimaeric tissue made up of cells from both species, e.g. cartilage or kidney tubules consisting of mixed chick and mouse cells. These experiments were carried out with relatively simple tissues at an early stage of embryonic development, but Weiss and James showed that even such complex organs as feathers would develop after dissociating the already-formed feather germs from 8-day chick embryos.

Dispersion and reaggregation of cells is not limited to experimental situations; it is a normal part of the slime mould's developmental history, and plays an important part not only in restorative processes in adult vertebrates, in regeneration and wound-healing, but in certain embryonic events, e.g. in the development of neural crest cells, to be described later. An unusual and striking example is found in the early embryogenesis of *Austrofundulus* and other annual fish, investigated by Wourms in 1972. In fish, cleavage results in the formation of a cap-shaped blastoderm at the animal pole of the egg, the rest of which consists of a large spherical yolk, from which the embryo develops. Annual fish live in water holes and swamps that dry up seasonally and, in order to avoid the effects of this dehydration upon the sensitive phases of gastrulation and neurulation which normally follow cleavage, the embryo responds to deterioration in the environment by entering a phase in which the deep blastoderm cells (the superficial cells produce an enveloping layer and a periblastic layer covering the yolk) disperse themselves by migration over the surface of the yolk as amoeboid cells. Some days later, the cells move back to form a reaggregate within which the normal embryonic processes are resumed in a further four or five days. Though the phenomenon is so dramatic, it appears to represent only an exaggeration and adaptation of processes involving changes in adhesivity and motility of the deep blastoderm cells which occur in the early embryogenesis of all fishes.

Cell movement in embryos

The major phase of embryonic cell movements is gastrulation, which follows cleavage and consists in a massive rearrangement of the blas-

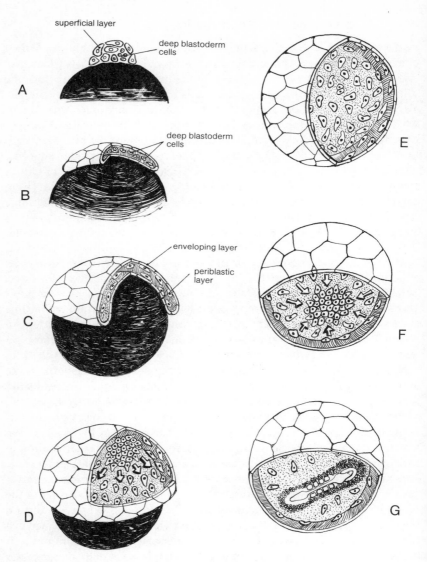

Figure 5.2 Cell movement in embryogenesis of the annual fish *Austrofundulus*.
A. The cap-shaped blastoderm.
B. and C. Blastoderm extension and division into superifical and deep blastoderm cells.
D. and E. Dispersal of amoeboid deep cells in adverse environmental conditions.
F. Reaggregation of these amoeboid cells.
G. Resumption of development and production of a normal embryo from the reaggregated cells.
Based upon J. P. Wourms.

tomeres to form three distinct germ layers: ectoderm at the surface, endoderm lining an interior cavity and mesoderm between. Thereafter, cell movements play a much less obvious part and only a few types of cell, e.g. the germ cells and derivatives of the neural crest, migrate over long distances. Development of the nervous system presents a special case, in which the neurones establish connections with effector or receptor organs, or with each other, over long distances by long cytoplasmic extensions (neurites), each drawn out by a migrating region at its tip which behaves like a minute moving cell, while the main cell body remains in its original position.

In the later stages cell movements become restricted to changes in shape and orientation which produce only very small shifts in relation to neighbouring cells, though these may still have important consequences. Ultimately, with the exception of the red and white blood cells (erythrocytes and leucocytes) whose function depends upon their ability to move in and around other tissues—passively in the first case and actively in the second—cells become immobilized as they become differentiated and establish firm connections with their neighbours. Yet in particular circumstances cells from late embryonic and adult tissues will detach themselves and resume an active motile and largely autonomous existence. Like the troops of Henry V before Harfleur, cells in tissues exist in a state of restraint, "like greyhounds in the slips, straining upon the start." This restraint may be released, as in wound-healing processes, or broken down as in cancer, when cells from a tumour which would in itself not be dangerous metastasize, i.e. begin active migration and colonization of other sites in the body. But the clearest examples occur in tissue culture, where cells of explanted tissue fragments move out onto a glass or plastic substratum and may, in some cases, be serially cultured as permanent lines of what are essentially independent proliferating unicellular organisms.

Cell movement *in vitro*

In these conditions the sociology of cells as interacting organisms may be studied with a view to discovering what elementary behaviour patterns form the basis of their much more highly organized activity in morphogenesis, the process through which complex structures are produced in the embryo. Since the physical basis of these activities in the properties of cell membranes and other subcellular structures can also be investigated in cultured cells, the way is open for analysis of morphogenesis in terms of molecular mechanisms and ultimately of their genetic control.

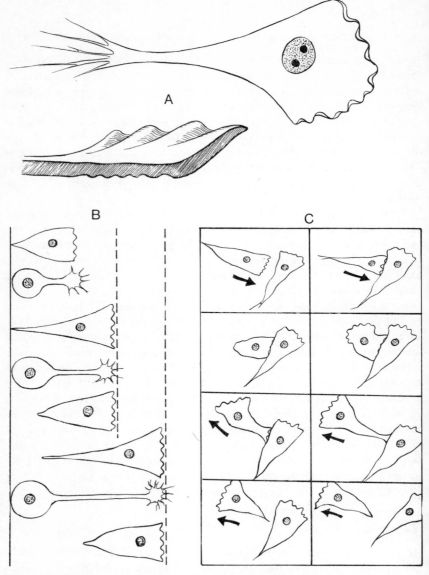

Figure 5.3 Cell movement and contact inhibition in cultured vertebrate fibroblasts.
A. A typical vertebrate fibroblast as it appears in culture on a glass or plastic substrate, and the ruffling lamellipodium at its leading edge in sectional view.
B. Comparison of the mode of movement in a fibroblast and in a neurone with an outgrowing nerve fibre, in which only the leading growth cone shows active locomotory activity (see chapter 17). Based upon N. K. Wessells.
C. Contact inhibition exhibited by fibroblasts from the corneal stroma, cultured on glass. Based upon J. B. L. Bard and E. D. Hay.

The study of cell behaviour in culture owes much to the groups led by P. Weiss, beginning in the 1930s, and M. Abercrombie, beginning in the 1950s. Weiss drew particular attention to the phenomenon of contact guidance, by which cells become orientated by alignment along fine structural inhomogeneities in the substratus, e.g microgrooves on a glass surface in culture or some sort of macromolecular exudate secreted by other cells in the embryo. The physical cues might be of macromolecular dimensions, and Rosenberg in 1963 showed that fibroblasts—cells which emerge from mesodermal structures in culture—orientated along grooves only 6 nm deep produced by building up monolayers of stearic acid on quartz. This form of control almost certainly operates in some embryonic situations, e.g in the growth of nerve fibres along muscle blocks and the growth of kidney ducts through the body, but it will at most provide a single-line track, leaving the direction the cell takes backwards or forwards along the track to be determined by other factors.

Abercrombie's studies were carried out on fibroblasts moving out from fragments of embryonic chick heart on a glass or plastic surface, using time-lapse cinéphotography. Each cell is characteristically flattened and polarized, with a broad leading lamellipodium and a trailing cytoplasmic region which gives it a roughly triangular shape. Ingram in 1969, photographing fibroblasts from the side, showed that the cell is attached to the substratum only at its trailing tip and at transient contact-points along its lamellipodium which displays a continual "ruffling" activity, caused by parts of the structure failing to attach to and lifting away from the glass surface. According to Ingram, movement occurs through extension of the lamellipodium a short distance, followed by its attachment to the substratum and subsequent contraction of the cytoplasm to draw the body of the cell after it, with corresponding release and reattachment at the trailing tip.

Fibroblasts emerging from tissue fragments in culture move radially away from the fragment and form monolayers, with no piling up or even over-lapping of adjacent cells. Abercrombie explained both these facts by reference to the phenomenon of contact inhibition which he observed when one fibroblast, moving at random, collided with another. When the lamellipodium of the colliding cell comes into contact with the other cell, its ruffling activity stops and it moves no further, so that no overlapping occurs; it adheres strongly to the membrane of the other cell, a new lamellipodium is formed at some other region of the cell periphery, and the cell moves off in that direction, stretching its adhesion to the other cell and finally breaking the attachment minutes or hours later. Thus, as long as

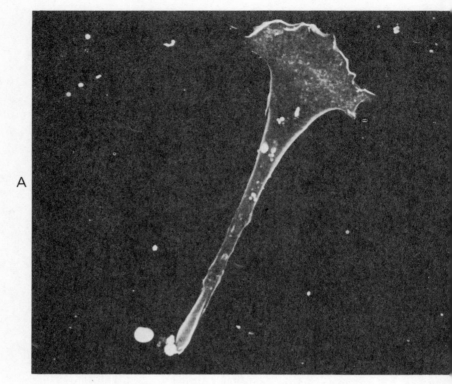

Figure 5.4 Mesenchymal cells of the chick somite sclerotome *in vitro* and in the embryo.
A. In culture, on plastic, showing typical fibroblast appearance with ruffling lamellipodium.
B. In culture, on plastic, showing extension of find filopodia from the leading edge.
C. In the embryo, shown by scanning electron micrography.

there is free space on the substratum, cells will move into it, and this will force them to move in general away from the explant to where free space exists.

How far does a cell's behaviour *in vitro* reflect its activity in the embryo?

In order to discover how far these observations of cells in highly artificial situations are relevant to actual developmental events, comparable studies must be made of real embryonic situations, but the problem of how to observe the behaviour of cells *in situ* has proved difficult to overcome. A successful attempt was made by Bard and Hay in 1975, using Nomarski microscope optics, which produce an extremely narrow depth of focus, and

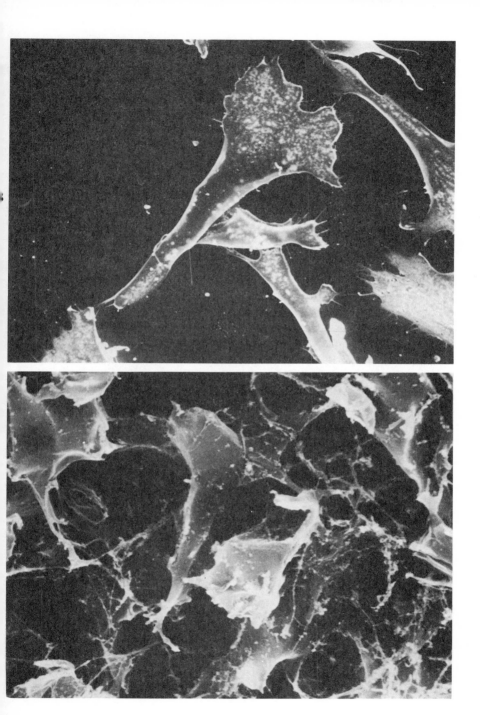

therefore the effect of optical sectioning through suitable preparations, in conjunction with a particularly favourable embryonic system—the cornea, the thin transparent membrane which covers the lens in the developing eye of the 6-day chick embryo. At this stage the cornea can be removed, with some of the peripheral ectoderm and mesenchyme but without the lens, and set up as an organ culture. It consists of two layers of cells—a surface epithelium, two cells thick, and an internal endothelium one cell thick— with, sandwiched between them a stroma, consisting chiefly of thin fibrils of the protein collagen, which is initially cell-free. Fibroblasts which have migrated to the peripheral mesenchyme from the neural crest, begin to move into the stroma from all sides, using the collagen fibrils as a substratum, and may be observed and filmed in much the same way as cells in a Petri dish.

Fragments of stroma into which some fibroblasts had already migrated were explanted into standard tissue culture conditions, on glass, and others into dishes in which the substratum consisted of a collagen gel, prepared by extraction from rat-tail tendon, in which each fragment was embedded. Fibroblasts migrated out from the fragments and, in the case of those moving onto glass, assumed the form and behaved in the way Abercrombie had described chick heart fibroblasts to do, i.e. flattened against the substratum, with a broad leading lamellipodium showing ruffling activity, and exhibiting contact inhibition on collision with another cell. Those migrating into the collagen gel presented a different appearance, elongated and clearly bipolar, with one or two long pseudopodia (thick cytoplasmic extensions) at the leading end, from which branched out fine filopodial cytoplasmic processes. The difference in form is clearly determined by the physical characteristics of the substratum, in one case a flat featureless surface and in the other a three-dimensional jungle of collagen fibrils, to which the fine filopodia probably attach directly. The collagen gel much more closely resembles the conditions within the embryo, especially within the corneal stroma, and in fact no differences were observed between fibroblasts in these two situations, either of form or behaviour.

In spite of these differences of form, fibroblasts in gel and in the stroma showed essential similarities in behaviour with fibroblasts on glass. In both, movement occurred by the leading process extending and attaching to the substratum, followed by contraction of the cell body towards it with a break of its adhesion to the substratum at the trailing end. More important, when one cell met another in the absence of other cells (a rather rare occurrence), the leading cell processes—lamellipodium in the one case and filopodia in the other—stopped moving and other cell processes took over

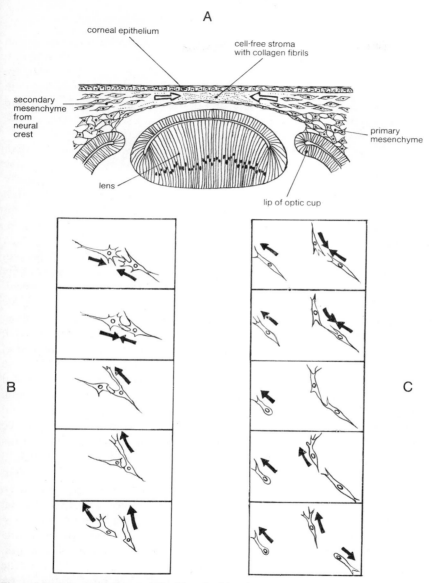

Figure 5.5 Behaviour of corneal fibroblasts in the embryo.
A. Section through the lens and neighbouring tissues, including the overlying corneum where neurectodermal fibroblasts are migrating into the collagenous stroma.
B. Corneal fibroblasts filmed *in situ* using Nomarski optics.
C. Corneal fibroblasts filmed in culture on a collagen-gel substratum.
Based upon J. B. L. Bard and E. D. Hay.

or arose distal to the point of collision, becoming motile and causing the cell to move off in another direction. These observations strongly support the hypothesis that contact inhibition of movement occurs not only in monolayers on a flat surface in culture, but also within the embryo, where it may be supposed to play an important part in controlling the distribution and arrangement of cells. In this case it provides a satisfying explanation for invasion of the corneal stroma, and Hay has suggested that one other example may be the movement of mesoderm cells away from the crowded primitive streak region in the chick gastrula.

It may also have an important negative function in confirming cells in their proper places within tissues once development is complete, and studies by Abercrombie and Ambrose on tumour cells suggested that in these cells contact inhibition was much reduced. This would provide a satisfying explanation of the characteristic invasiveness of neoplastic (i.e. malignant) cells, but subsequent work has thrown some doubt on the interpretation of the original observations. What was thought to be overlapping of tumour cells in culture may in fact be due to "underlapping", resulting from the cells' failure to adhere to the substratum as closely as do normal cells, and consequently allowing other cells to move beneath them. This and other questions regarding the special nature of cell contacts and interactions in cancer cells remain open and are of immense importance.

Armstrong and Armstrong showed in 1973 that cells are in any case not necessarily completely immobilized in the middle of tissues, by fusing fragments of embryonic kidney from chick and quail, whose cells can be distinguished by their staining properties. They showed that cells from one fragment were to be found deep in the other fragment, indicating movement from one to another. The cell sorting exhibited in Moscona's experiments implies some sort of movement of cells within a mass of other cells, and direct observations by Garrod in 1973 on sorting in monolayers suggest that this is due to active cell locomotion rather than passive displacement by physical forces. The field is one of great complexity and is the subject of much current research. Problems of cell movement are almost inextricably linked to problems of cell adhesion, and on this subject there is an enormous diversity of hypotheses, none of which is yet established with certainty, ranging from the quantitative differential-adhesion concept of M. S. Steinberg to the cell ligand hypothesis of Moscona, involving specific molecular but dynamic connections between one cell and another. A particularly promising approach to the problem of control of cell adhesion in morphogenetic systems has been developed by

Curtis, based upon his discovery with van de Vyver in 1971 of soluble factors produced by sponge cells which increase the adhesiveness of homologous cells but decrease that of heterologous strains. In view of their implications regarding the nature of cell interactions in normal and pathological developmental processes, the importance of these studies on the cell surface, which cannot be dealt with here, cannot be overemphasized.

CHAPTER SIX

MORPHOGENETIC MOVEMENTS IN EARLY EMBRYOGENESIS

Cell rearrangements and presumptive-fate maps

WE HAVE SEEN THAT THROUGH CLEAVAGE, IN ONE WAY OR ANOTHER, A hollow blastula is formed, with walls which are sometimes one cell thick, as in echinoderms and insects, and sometimes several as in amphibians, in which the blastocoele roof is thin, usually two micromeres in depth, but in which the floor, consisting of the much larger macromeres, is bulky and multilayered. There follows a period of extensive and beautifully co-ordinated rearrangements of cells, either individually or as sheets, known as *morphogenetic movements* since they define the basic form of the embryo. Gastrulation, neurulation and the formation of extra-embryonic membranes are the chief categories, but one flows so smoothly into another that it is sometimes difficult to distinguish them completely.

It is often important for the interpretation of embryonic experiments to know what each region of the blastula will give rise to in the fully developed embryo, and for this purpose, and in order to follow the morphogenetic movements, Vogt in the 1920s devised a method for constructing fate maps by marking patches on the amphibian blastula with harmless (vital) dyes, such as neutral red or Nile blue, and tracing these patches to their positions in the embryo at later stages. More precise methods have since been introduced, e.g. labelling cells with a radioactive marker such as tritiated thymidine and grafting them back into their normal sites, or using cells of a different species which have different nuclear staining properties; and for insects, where these direct methods cannot be used, an ingenious genetic mapping technique has been developed (see chapter 15).

In embryos with mosaic developments the relationship between any region of the blastula and its subsequent development is invariable, but in regulative embryos the map will only apply if normal spatial relations are preserved and the fate of the different parts is therefore said to be *presumptive* or *prospective*.

Gastrulation

In the next stage of development, known as *gastrulation*, the cells of the blastula are rearranged to produce a more complex embryo consisting of three germ layers—the *ectoderm, mesoderm* and *endoderm*, each of which will give rise to particular structures in development: the ectoderm to the epidermis and nervous system; the endoderm to the gut wall and its glandular derivatives; the mesoderm to all other internal structure—the blood system, the muscles, the skeleton and so on. At the end of gastrulation, only the ectoderm remains at the surface of the embryo. The distinction between the germ layers is not so rigid as was formerly thought (see discussion of the neural crest in chapter 7) but, apart from being useful for descriptive purposes, the division into these three basic sheets of cells does appear to constitute a fundamental feature of early embryonic organization. Gastrulation brings these layers into relation with each other, so that in mosaic embryos the appropriate parts are ready to continue developing in proper juxtaposition, and in regulation embryos appropriate inductive interactions can occur between one layer and another.

Gastrulation in echinoderm embryos

Gastrulation movements are simplest in embryos whose cells contain very little yolk, such as those of echinoderms. The latter constitute especially good material, because they are so small and transparent that it is possible to follow the activity of individual cells within the embryo using time-lapse cinéphotography, as Gustafson, with Kinnander and Wolpert, did in the 1960s, using the sea urchin *Psammechinus*.

The echinoderm blastula is spherical and made up of about 1000 ciliated cells attached to a distinct surface hyaline plasma layer. After about 24 hours of free-swimming life, the cells resorb their cilia and the blastula becomes flattened at its vegetal pole, forming the vegetal plate. Here about 40 cells show pulsatory activity and round up, breaking contact with the surface layer and their neighbours, and squeezing into the blastocoele. Here they move around the inner surface of the blastula cells, making apparently random exploratory movements but eventually taking up a position in a doughnut-shaped configuration around the edge of the vegetal plate; possibly there is some matching of points of attachment to particular attachment sites. Locomotion occurs by extension of several long branched filopodia, which attach by their tips to the junctions between the blastula

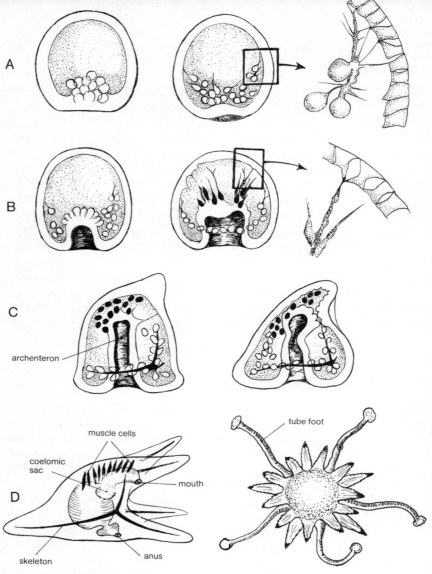

Figure 6.1 Gastrulation and development of the pluteus larva in the sea urchin *Psammechinus*.
  A. Appearance and migration of the primary mesenchyme cells, which later secrete the skeleton of the pluteus.
  B. Appearance and behaviour of the secondary mesenchyme cells (coloured black), which first haul the archenteron up to the animal pole, then produce the larval muscles (C).
  C. and D. Stages in formation of the bilaterally symmetrical pluteus larva and its transformation into the radially symmetrical adult form.
Based upon T. Gustafson and L. Wolpert.

cells, and subsequent contraction of these filopodia, in which Tilney and Gibbons have demonstrated the presence of orientated microtubules. These cells are the primary mesenchyme cells, which produce the skeleton of the pluteus larva; Okazaki showed in 1960 that filopodia from several cells fuse, forming a syncytial cable in which the spicules of the skeleton are secreted, in a process which recalls the integrated activity of spicule-secreting cells in sponges.

While the primary mesenchyme cells are taking up their position, the remaining cells of the vegetal plate begin pulsating and rounding up on their inner borders, but this time without detaching from each other or from the surface layer, so that the mechanical effect is to produce a hemispherical impushing or invagination of the whole plate into the blastocoele. This is followed by a second phase of gastrulation in which the invagination is extended across the blastocoele, forming a deep pit known as the *archenteron* or primitive gut. This elongation is brought about by cells at this end throwing out long filopodia which attach by their tips to the blastula wall, again to the junctions between ectodermal cells, and then shorten, hauling the archenteron up to the animal pole of the blastula. These cells eventually withdraw from the archenteron wall and become secondary mesenchyme cells within the blastocoele, mostly producing the larval musculature. Finally the tip of the archenteron bends forward and fuses with the ectoderm, and by perforation forms an oral opening, while an outpocketing (evagination) just behind the tip produces the system of larval coelomic vesicles. Towards the end of larval life a small echinoid rudiment is formed by invagination into a cavity which becomes closed, rather as the amniotic cavity is formed in amniote vertebrates; and this rudiment grows within the larva around the larval organs until, at metamorphosis, the free-swimming larval form is replaced by the hemispherical form of the bottom-living slow-moving adult sea urchin.

Gastrulation in amphibians

In the protochordate *Amphioxus*, whose embryological development is of a very simple and probably primitive chordate type, gastrulation occurs essentially as in echinoderms, by a hemispherical invagination appearing in a hollow spherical blastula whose wall is only one cell thick. In amphibians the process has essentially to bring about the same result, but by a mode of infolding which can be accomplished in spite of the vegetal half of the blastula consisting of relatively inert yolky macromeres. Amphibian gastrulation has been studied in both urodeles (with tailed adults—newts and axolotls) and anurans (with tailless adults—frogs and

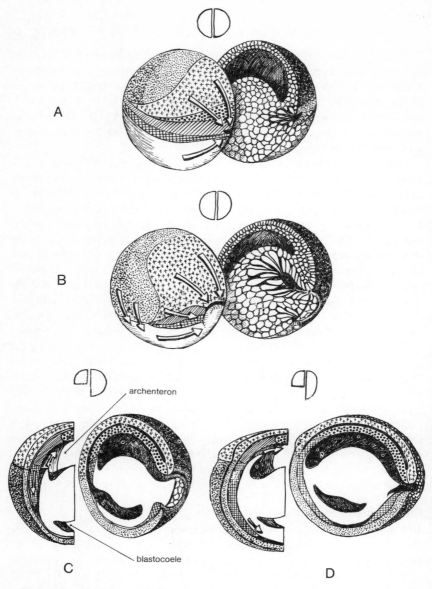

Figure 6.2 Gastrulation in amphibians.
A. An early gastrula, one half showing formation of the dorsal blastopore lip in the grey crescent area, with flask cells. The other half shows the presumptive fate of the various regions.
B. A midgastrula, with some surface material deeply invaginated and the blastopore now horseshoe-shaped.

toads). The various regions of the blastula are more clearly distinguished by pigment differences in anurans—a black-pigmented animal region of micromeres, the presumptive ectoderm; an unpigmented vegetal region of macromeres, the presumptive endoderm; and a roughly equatorial band between them which is grey and approximates to the presumptive mesoderm; within this band lies the grey crescent of the fertilized egg.

Gastrulation begins with the formation of a crescent-shaped groove of invaginating cells (the dorsal blastopore lip) in the region of the grey crescent, just below the equator of the blastula and defining what will become the dorsal side of the embryo. Along this groove cells roll over by a process known as *involution* into the interior, to be followed by other blastula cells which stream towards the blastopore before themselves rolling in after the others. Continuous with this process, the presumptive ectoderm cells grow down over the macromeres, in a process called *epiboly*. As more of the blastula cells move in, so the crescent of the blastopore lip becomes extended to form lateral lips as well, producing a horseshoe-shaped groove, and finally is completed ventrally to form a circle, with the yolky macromeres which are still not enclosed protruding as a white yolk plug, surrounded by black ectoderm; these macromeres disappear into the interior and only a small blastopore opening remains. Because they are bulky and slow-moving, the macromeres give the impression of being inert and covered by the active presumptive ectoderm, but in fact they also move in at the blastopore, though more slowly.

C. A late gastrula showing the blastocoele almost obliterated and the appearance in its place of the archenteron. The blastopore is now a complete circle, plugged with the yolk plug—macromeres which are still visible externally; otherwise the material at the surface is entirely ectodermal.

D. An early neurula, showing beginning of elongation and the delineation of the neural plate, i.e. the visible division of neural ectoderm from the rest. Dorsalward movements of the endoderm and ventralward movements of the mesoderm complete the endodermal and mesodermal germ layers.

Based partly upon J. Holtfreter.

Presumptive fates:

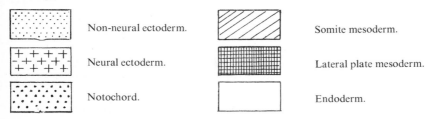

On the inside the invaginated dorsal material (the presumptive mesoderm) pushes forward hard against the ectoderm, forming the roof of the archenteron; the cells in the mid-line will form the axial notochord, and are sometimes distinguished as chorda cells, but the whole region forms a continuous sheet—the *chordamesoderm*. The macromeres now form the floor of the archenteron, and the lateral edges of the chordamesoderm grow down between them and the presumptive endoderm, then fuse ventrally to give the three-layered embryo. The endoderm cells, which at first formed only the floor of the archenteron, move upwards at each side and over to form a roof, coming to enclose the archenteron completely. At this stage the presumptive ectoderm covers the whole of the embryo—now called a *gastrula*—and becomes stretched in epiboly to give a single layer of cells in urodeles and a double layer in anurans.

Most modern work on the mechanics of gastrulation in amphibia stems from the work of Holtfreter in the 1930s and 40s. Holtfreter drew particular attention to the so-called *flask cells*, noted by Rhumbler in 1902, formed in the vicinity of the blastopore lip, and to the related importance of what he called the *surface coat*, comparable to the surface layer of the echinoderm blastula; this he believed to constitute a distinct entity, but it may be simply the surface layer of the blastula cells, cohering firmly from one cell to another. As cells approach the blastopore lip, they become lengthened and squeeze actively into the interior, retaining a connection with the surface coat by a cytoplasmic stalk—the neck of the flask—which becomes more and more attenuated as the cell moves inwards; if they were not connected to the surface coat, they would simply pass in as individuals but, because they are connected to it, they produce the blastoporal invagination. All this Holtfreter demonstrated by observations on intact and dissected embryos, and in combinations of isolated fragments of presumptive ectoderm and endoderm. His studies on the selective affinities of cells of the three germ-layers after they had been segregated led very largely to the whole field of research into cell adhesion as a mechanism of embryonic development.

Reviewing the whole field of morphogenetic movements in 1976, Trinkaus pointed out that flask cells can only account for the first (invagination) phase of gastrulation and not for the massive involution which follows as presumptive endoderm and mesoderm cells move interiorly to form the archenteron, and that a number of authors believe that an important role is played by active locomotion of presumptive mesoderm cells. Nakatsuji in 1975, using scanning electron microscopy, showed that in the urodele *Cynops* some invaginating mesoderm cells flatten against the blastocoele wall and probably, by means of filopodial activity, migrate along its inner surface; Keller and Schoenwolf reported in 1976 that in *Xenopus* prospective head mesoderm cells become locomotory in the late blastula and move inwards between the ectoderm and endoderm layers, carrying the presumptive endoderm cells along as passengers.

Finally, it should be noted that although it is now generally accepted that most morphogenetic movements result from integrated but autonomous activities of individual cells, several observers, including Holtfreter, have reported the puzzling phenomenon of "pseudogastrulation" in unfertilized amphibian eggs, in which crude but distinctly gastrulation-like changes of form occur without subdivision into cells. The most striking report was by Smith and Ecker in 1970; by treating eggs removed from the ovary of the frog *Rana pipiens* with progesterone, a blastopore-like groove and then a "yolk-plug" appeared about $2\frac{1}{2}$ days after removal to a simple balanced salt solution. In insects (see p. 76) supracellular cytoplasmic movements play an important part in normal embryogenesis, and it seems that the potential for some similar programmed cytoplasmic movements exists in amphibians, though whether or not they have any function in normal development is unknown.

Early morphogenetic movements in the avian embryo

On account of its easy availability and its accessibility within the egg shell, the chick embryo has been a favourite subject for embryological studies since Aristotle, in the 3rd century BC, described the appearance of the beating heart at the third day of incubation. But these advantages do not extend to the very earliest stages, since development begins immediately after fertilization in the oviduct of the hen and continues through the commencement of gastrulation and the formation of the primitive streak, so that a good deal remains to be discovered.

Moreover, the quantity of yolk in the avian egg is so disproportionately great that cleavage, blastula formation and gastrulation assume forms which appear to be quite different from the corresponding processes in amphibians. However, in their basic features there are essential similarities, and it is possible to see how the comparatively simpler processes have been modified in the course of evolution to fit a situation in which enlargement of the yolk has reduced the cytoplasm to a mere cap at the animal pole of an enormously large egg cell. Thus the process of cleavage, in which cytoplasmic division is at first by furrows produced by downward extensions of the egg cell membrane, leaving the blastomeres open to the yolk on their vegetal sides, is clearly an adaptation of the massive amount

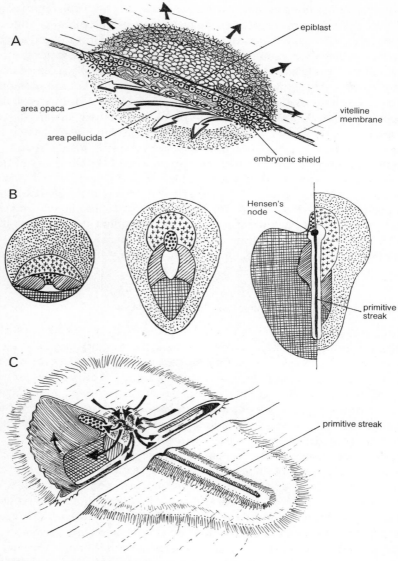

Figure 6.3 Gastrulation in the avian embryo.
A. Early blastoderm, expanding radially; with division into area opaca and area pellucida, the embryonic shield from which cells fan out into the area opaca.
B. Fate maps from the beginning of gastrulation to the formation of the primitive streak. (The surface layer has been removed from the left-hand side in the last stage).

of yolk and is paralleled in the formation of the syncytial blastoderm of insects.

Most descriptions of the early events are based upon the work of Patterson on the pigeon in 1902, but recent advances have been reviewed by Bellairs in 1971. In later stages of cleavage the central cells of the blastodisc become cut off on all sides but, for some time, at the edge (the periblast) there is a syncytial region where cells still remain in open communication. Further cleavage produces a blastoderm—a round disc, 5–6 cells deep in the centre (the epiblast) but only 1 or 2 cells deep at the periphery. The beginning of gastrulation is marked by the division of the blastoderm into a peripheral *area opaca*, which adheres closely to the yolk and appears opaque in transmitted light, and a clear central *area pellucida* in which the embryo is formed, appearing first as the primitive streak. The epiblast forming the area pellucida is now only 2 or 3 cells deep, with a slight gap between it and the underlying yolk corresponding to the blastocoele; at the posterior end of the area pellucida is a small opaque, rather thicker, area called the *embryonic shield*. The upper surface of the marginal cells of the area opaca adhere to the undersurface of the vitelline membrane of the egg and spread over the yolk by moving over it in the process of epiboly, eventually enclosing all of it in a yolk sac. Downie and Pegrum showed in 1971 that all yolk-sac cells are capable of moving over vitelline membrane in tissue culture, but only the marginal cells are capable of organizing together in the particular way necessary to produce the epibolic spreading. Mitosis evidently plays no part in this process, since it goes on in the presence of mitotic inhibitors, and it is produced entirely by the marginal cells, moving by throwing out extremely long filopodia—up to 500 μm—and contracting them, pulling the more central cells along passively behind them.

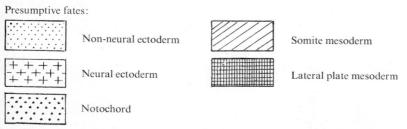

Presumptive fates: Non-neural ectoderm, Neural ectoderm, Notochord, Somite mesoderm, Lateral plate mesoderm

C. Partly sectioned view of the primitive streak stage, showing invagination of endodermal and mesodermal material, with some material at the anterior end of the streak (Hensen's node) moving anteriorly to form the notochord.
Based upon various sources, summarized by R. Bellairs.

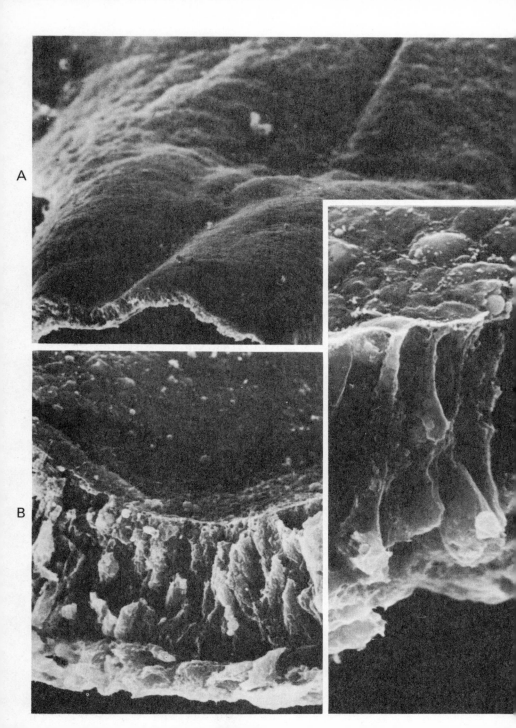

A rule dating back to Von Baer in 1828 states that if the egg is held with its pointed end to the right, the future antero-posterior axis of the embryo will be at right angles to the long axis of the egg, and the head will point away from the observer. Generally speaking this rule holds good, indicating an early determination of the axis, but there is a very high regulative capacity at this stage. Lütz in 1949 showed that in the duck embryo, which hatches at an earlier developmental stage than the hen, if the blastoderm is cut into four parts, each one will regulate to give a normal embryo; the posterior embryo from a halved blastoderm tends to have the original orientation, and the anterior one will be orientated at random. Slowing down the rate or delaying the start of the development, e.g. by cooling, often leads to production of multiple embryos. Development in the absence of fertilization, as established in a parthenogenetic strain of turkeys by Olsen in the 1960s, in which there is a delay of 2–3 days in the renewal of development after the beginning of incubation, causes polyembryony. Two to eight embryos may be produced, with variation of size and location on the blastoderm suggesting that they originate at different times and positions. This effect is not limited to avian embryos, and a similar developmental lag with subsequent polyembryony occurs regularly in parthenogenetic insects such as the midge *Miastor*.

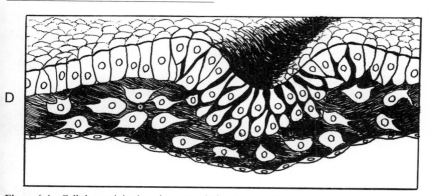

Figure 6.4 Cellular activity in avian gastrulation (see p. 74).
A, B and C. Scanning electron-micrographs of a chick blastoderm, sectioned at the primitive streak, showing Hensen's node in A, with the outlines of somites forming in the invaginated mesoderm anterior to it. The appearance of the epiblast cells as they invaginate is shown at higher magnification in B, and in greater detail in C.
D. Diagrammatic illustration of the above.
Based upon E. D. Hay.

Gastrulation in the avian embryo

The most fundamental difference between amphibian and avian gastrulation is that no archenteron is formed in bird embryos; the origin of the endoderm has long been a matter of confusion and uncertainty. The earliest suggestion, by Patterson in 1909, that there is a temporary blastopore at the posterior edge of the blastoderm, is no longer accepted. In 1960 Spratt and Haas showed that there was a movement of cells from the lower side of the embryonic shield, which fans out as it advances forwards over the yolk, but that these cells move into the area opaca where they produce the extra-embryonic endoderm (hypoblast) and possibly mesoderm of the extra-embryonic blood system. The primordial germ cells were shown by Dubois in 1962 to originate in the posterior margin of the unincubated blastodisc, and to be then carried forward to their anterior position in this fanwise movement.

The problem appears to have been solved in 1962 by Vakaet and Marcel, who discovered that removal of the embryonic endoderm (endoblast) at an early stage is followed by formation of a new one, which can be filmed using time-lapse photography. They showed that the new endoderm, and presumably the original one, arose by migration of cells from the primitive streak and out to both sides to form a floor to the "blastocoele," so that, if the primitive streak is thought of as a blastopore elongated in the antero-posterior axis, gastrulation in avian and amphibian embryos is not fundamentally different. At about the same time, mesoderm cells also begin to migrate out from the primitive streak, moving between the epiblast and the endoblast, but the latter is completed long before the mesoderm. As cells invaginate they elongate and become flask-shaped as in amphibian gastrulation, but in this case there is no surface coat; as they move inside, cells break their attachment to other surface cells and consequently only a shallow groove is produced along the mid-line of the primitive streak. The cells of the streak are replaced by others which stream in from the epiblast, and the movements of the epiblast cells and their presumptive fates have been mapped by Rosenquist, using tritiated-thymidine-labelled grafts in unlabelled blastoderms. At this stage the epiblast is a single-layered columnar epithelium (at the streak it is multilayered), the endoblast a singly-layered squamous (pavement-like) epithelium, and the mesoderm a loosely organized mesenchyme.

Morphogenetic movements in insect embryos

The embryonic development of insects is as complex as any in vertebrates, and includes the same types of morphogenetic movements—of gastru-

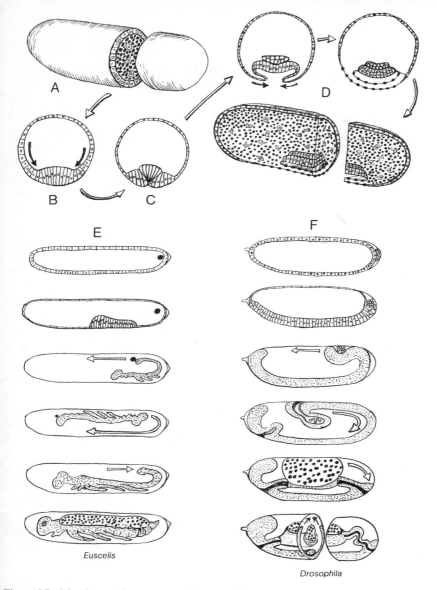

Figure 6.5 Morphogenetic movements in insect embryos.
A–D. Generalized diagram to illustrate development in a typical insect egg from the cellular blastoderm stage (A), showing formation of the ventral germ band (B), mesoderm (C), the extraembryonic membranes (D). The endoderm appears as two rudiments, an anterior and a posterior one, pushed into the interior by invagination of ectodermal stomodaeal and proctodaeal rudiments.
E. Development in the homopteran *Euscelis*, showing blastokinesis. Based upon K. Sander.
F. Development in the dipteran *Drosophila*, showing modified blastokinesis and illustrating the way in which the pole cells are taken into the interior of the embryo, formation of stomodaeal and proctodaeal rudiments, and dorsal closure.
Based upon D. A. Ede and S. J. Counce.

lation, neurulation, formation of extraembryonic membranes—and some more beside; they have been well reviewed by Sander in 1976. Cleavage produces a single-layered blastoderm enclosing the ooplasm—a massive yolky cytoplasm whose cellular structure is often not clear, which is capable of controlled contractions and other movements. It is probably these which lead to an unevenness in distribution of the blastoderm cells, which become multilayered along a germ band, from which the embryo will be produced; the rest of the blastoderm forms extraembyronic membranes. Later, the sides of the germ band move up and over the ooplasm in a dorsal closure movement which encloses the yolk in the endoderm and completes the embryo dorsally.

Insect embryos are transparent so that time-lapse filming can be used in the analysis of their movements, as for *Drosophila* by Ede and Counce in 1956, and for other insects by other workers subsequently. The most striking movement, which is unique to insects, is called *blastokinesis*, and it is best shown in hemimetabolous insects such as *Euscelis*, in which the germ band occupies a relatively small part of the egg. Here the whole of the germ band moves up and over to the dorsal side of the egg and then forward, rotating on its long axis also in some insects; then, after a further period of development in that position, the movement is reversed and the embryo returns to its original position on the ventral side. No convincing reason has been produced for the existence of this movement in insects, and it remains a puzzling oddity. In *Drosophila*, where the germ band occupies the whole length of the ventral side of the egg, blastokinesis takes the form of an elongation of the germ band, the posterior tip of which moves over and forwards, producing a U-shape: then, after some time, shortening to resume its previous position. In time-lapse film this movement appears as sudden and dramatic, the posterior tip scooping up the germ cells and carrying them forward to a position just behind the head. The mechanism may be partly simple crawling of posterior germ band cells on the inner surface of the extraembryonic blastoderm, but the surge of activity suggests something more; in 1972 Vollmar showed that in the cricket pulsations of the ooplasm are also involved.

CHAPTER SEVEN

NEURULATION AND THE DEVELOPMENT OF THE EMBRYONIC AXIS IN VERTEBRATES

AT THE END OF GASTRULATION ALL OF THE PRESUMPTIVE ECTODERM remains at the surface of the embryo, but a large area of it has still to be rearranged to form the neural tube—the rudiment of the brain and spinal cord, and the source of all nerve cells (neurones) in the body; this process is termed *neurulation*. In the course of it, the vertebrate embryo also lengthens in the antero-posterior axis, and the mesoderm becomes divided up to form an axial rod (the *notochord*), the paraxial plate on each side which breaks into a series of paired segmented somites and, lateral to these, the lateral plate which splits to give a central cavity (the *coelom*) and the tissues lining the body wall to the outside and the gut to the inside.

Neurulation in amphibia

In the amphibian gastrula, the presumptive neural ectoderm forms a nearly hemispherical sheet, one cell thick in urodeles and two cells thick in anurans, on the dorsal surface. As neurulation begins, this sheet becomes flattened, deformed to produce a key-hole shape, and defined by a ridge around its periphery to give a clear neural plate in which the ectoderm is thicker than in the surrounding presumptive epidermis, though there is no increase in the number of cell layers. An infolding occurs in the neural plate, whose lateral margins meet and then fuse in the dorsal mid-line, producing first of all a trough and then a tube (the neural tube) whose lumen is much wider anteriorly on account of the shape of the original plate, and which remains open posteriorly by a small neuropore. From the first appearance of the folds to their fusion in the midline takes only a few hours.

The forces which produce neurulation have been much investigated, and this particular morphogenetic movement is the one in which deformation in a sheet of cells, changes in the form of individual cells, and their relation to subcellular structures, have been most clearly worked out, particularly by

Baker and Schroeder in 1967 for the South African clawed toad *Xenopus*. The first cellular indication occurs along the median line in the superficial layer of the 2-layered neural ectoderm, where cells change from being cuboidal to elongated—along the floor of a shallow neural groove, and then to flask-shaped—precisely and symmetrically located at the bottom of a deep neural groove; wherever the walls of the groove are convex, the cells of the superficial layer are flask-shaped and, wherever they are straight, the cells become flattened. In the deep layer of neural ectoderm, the originally cuboidal cells elongate but remain columnar; those immediately beneath the flask cells are pushed to the side, so that a thick lateral placode of columnar cells is formed on each side of the neural groove. In the most lateral region of the deep layer, the cells become rounded and detached from their neighbours, and in the later stages of neurulation move away from the neural plate, migrating for long distances and differentiating as a wide variety of cell types; these are the cells of the neural crest.

Microtubules are straight, unbranched and hollow cylindrical organelles which occur in a very wide variety of cells, where they may be aligned in parallel arrays, roughly orientated in line, or orientated at random; they are assembled and resorbed rapidly. Frequently they serve as passive cytoskeletal supports, but in 1964 Byers and Porter showed that in the very rapidly elongating cells of the developing chick lens about 100 microtubules appeared in each cell, aligned along the elongating axis, and proposed that the assembly of these ordered arrays of microtubules was responsible for the cellular elongation. In 1970 Pearce and Zwann confirmed that this was the case by treating lens cells with colcemid, which selectively destroys microtubules, and finding this led to loss of the distinctive cell shape. Baker and Schroeder set out to look for orientated microtubules in neural plate cells, and found that they did indeed occur, especially in the flask cells, where there were about 150 in each cell during or after the period of obvious elongation, but not at other times.

Elongation would not in itself produce infolding, so that some other force is required. Cloney had shown in 1966, investigating the sudden retraction of the tail in the metamorphosing ascidian (sea-squirt) larva, that the epidermal cell contraction which produced it was brought about by very rapid assembly and/or alignment of fine microfilaments (now known to consist of an actin-like protein) in the cytoplasm, and he suggested that these organelles might play a part in many developmental processes, including neurulation. Baker and Schroeder, following up this lead, found that in the *Xenopus* neurula microfilaments are found exclusively in the flask cells, but only at precisely the stage of infolding, and arranged in large

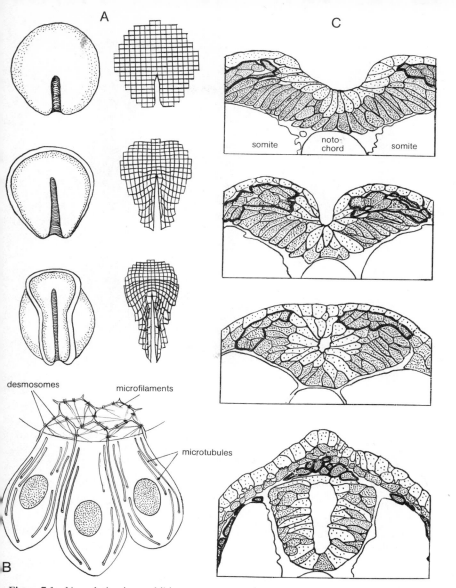

Figure 7.1 Neurulation in amphibians.
A. Changes in shape of the neural plate in neurulation and computer simulations of this process showing how they may be accounted for by changes in shape and position of the neural cells. Based upon A. G. Jacobson and R. Gordon.
B. Microtubules producing elongation and microfilaments producing a purse-string effect in neural cells. Based upon B. Burnside.
C. Neurulation in *Xenopus*. The superficial layer of the ectodermal is lightly stippled and the deep layer heavily stippled. Neural-crest cells are enclosed by a heavy line; in the bottom picture they are seen dispersing laterally, moving out between the ectoderm and the somite; others will move out between the somite and the neural tube. Based upon P. C. Baker and T. E. Schroeder.

80 NEURULATION AND THE DEVELOPMENT OF THE EMBRYONIC AXIS

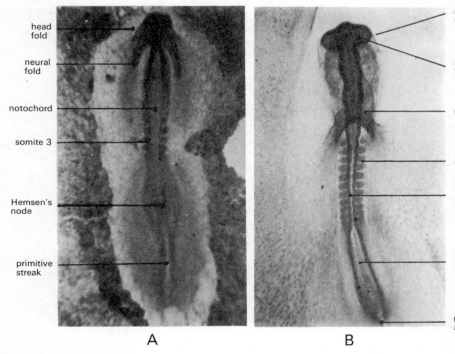

Figure 7.2 Gastrulation, neurulation and somite formation in the avian embryo (see p. 82).
A. Fowl embryo of approximately 26 hours incubation showing shortening primitive streak and regressing Hensen's node. Anteriorly, somite formation and neurulation has begun, and the head region is lifting off the surface of the blastoderm by formation of the head fold.
B. A later stage (approximately 36 hours), with node regression complete, 11 somites formed, and development of the brain and heart well advanced.

skeins around the apical perimeter of each flask cell; it appears that the microfilaments function as an intracellular purse-string, and the manner in which the cell membrane is thrown into folds, with the microfilaments positioned at the level of smallest cell dimension and anchored to the cell membrane at desmosomal junctions, support this. The role of microfilaments in other developmental systems has subsequently been explored by Wessells and others, and found to be as widespread as Cloney originally suggested.

In amphibian neurulation then (in 1971 Burnside showed that the same held for urodeles as for anurans), the shape-changes of the neurula cells are

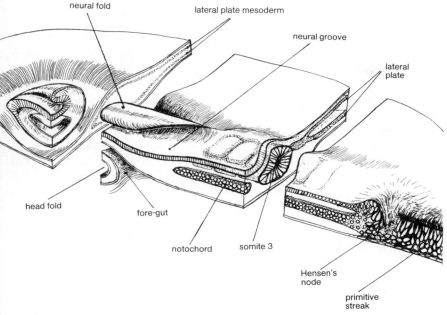

C. Diagrammatic illustration of the embryo shown in A.

active and autonomous rather than imposed by forces external to them, but the morphogenesis of the neural tube is still not entirely accounted for, even in mechanical terms. In 1976 Jacobson and Gordon analyzed this problem in a combined experimental and computer simulation study of changes in shape of the urodele neural plate (incidently, proposing a cellular basis for D'Arcy Thompson-style transformations—see chapter 13). They showed that to produce the observed changes there must be a regionally programmed shrinkage of the surface of the neural plate, which at the cellular level would be produced by the microtubule-microfilament mechanism described above, but also a displacement of the whole sheet by elongation in the antero-posterior axis, and that this might either be another autonomous activity of the neural plate cells in the mid-line, or be the result of their close association with the elongating notochord which underlies them.

Gastrulation, neurulation and axial organization in birds

In avian embryos it is not possible to describe these events entirely separately, because they flow into one another sequentially, each being most advanced at the anterior end of the embryo; thus in an egg which has been incubated for one day neurulation and somite formation (beginning with the demarcation of cuboidal blocks—segmentally arranged—in the primary mesenchyme) will be going on anteriorly, while gastrulation is still proceeding posteriorly; there is, as it were, a wave of embryonic development which passes backwards down the embryonic axis.

At the anterior end of the fully-developed primitive streak is a region in which invagination is particularly active, known as Hensen's node, with a depression in the middle of it corresponding to the anterior end of the primitive groove. At Hensen's node, not only do cells move into the interior and then outwards laterally between the epiblast and endoblast to give the mesenchymal mesoderm, but other cells move anteriorly to form the notochord. As the notochord is produced in this way, Hensen's node (i.e. the region in which notochordal cells are invaginating) moves back so that it is always situated at the most posterior region of the notochord, feeding more cells in to add to its length from behind. As Hensen's node moves back (regresses), so the primitive streak is shortened, and in front of it neurulation and somitogenesis begin, until eventually they are occurring along the whole length of the embryo, and the streak and the node have disappeared. Not only does segmentation follow node regression in time and space, but the work of Bellairs in 1963 suggests that the regression movements actually initiate somitogenesis, since removal of the influence of other tissues previously suggested as playing this role (e.g. the hypothetical somite centres postulated by Spratt in 1955) did not inhibit somite formation unless regression of the streak was itself prevented. It appears that the shearing force it generates may lead to cell surface changes which directly or indirectly induce segmentation. This idea received support in 1974 from the work of Lipton and Jacobson, who showed that if a portion of chick blastoderm which does not contain either streak or node is physically split with a needle, somites develop in this region within 6 hours, but simultaneously rather than in sequence, presumably reflecting the difference in speed between the swift cut with the needle and the slow regression of the node. They also showed, in time-lapse film, how the retreating node does clearly divide the mesodermal mesenchyme into two halves. These artificially induced somites became disorganized after a time, probably because of the absence of notochord, to which somites normally make connections which have a stabilizing effect. Lipton and Jacobson believe, however, that contact with neural plate is also necessary for somite development.

Cell contacts and cell communication in early chick embryos

Many workers have been attracted to the early stages of chick embryogenesis as material for investigations on cell junctions in the developing embryo, largely inspired by the electron-microscope study of epithelial tissues in 1963 by Farquhar and Palade, who introduced a standard terminology for the types of junctions they found. Most influential among these was the study of Trelstad, Hay and Revel in 1967 on cell contacts in early morphogenesis, but technical advances have brought many fresh discoveries, especially by Revel, Chang and Yip and by Sanders in 1973, and Bellairs, Breathnach and Gross in 1975. The original terminology has been considerably simplified, and three chief types of junction are recognized in the early chick embryo—desmosomes (large complex junctions, concerned with binding one cell firmly to another, and not present in great numbers), tight junctions and gap junctions. A clear distinction between the last two has only been possible since the introduction of the freeze-fracture replication technique in the 1970s, which has made it possible to visualize the appearance of these junctions in surface view rather than only in section. Tight junctions are long thin regions of union between two cell membranes, probably acting chiefly as permeability barriers between cells, and found widely in the chick embryo. Gap junctions are areas of close contact between adjacent cell membranes which are separated in this case by an extremely narrow gap of 2–3 nm; freeze-fracture pictures show that they form extensive patches composed of particles packed in hexagonal arrays. They are regions of increased permeability to movement of small ions from cell to cell, and therefore of low electrical resistance, but also to larger (but not very large) molecules, e.g. fluorescein (mol. wt. 330), a tracer which can be injected into one cell and its progress into other cells, which takes place only by way of the junctions, subsequently followed. Recognition of the extent to which metabolites may pass from one cell to another by way of gap junctions has followed the discovery of Subak-Sharpe, Burke and Pitts in 1966 that metabolic co-operation occurred between mutant and normal hamster cells in tissue culture. The mutant cells were incapable of incorporating ^{3}H-hypoxanthine into cellular RNA and DNA owing to the absence of the enzyme HGPRT (hypoxanthine: guanine phosphoribosyl transferase) but

Figure 7.3 Gap junctions between cell membranes.
A. A gap junction between cells of the notochord in the chick embryo, shown by the freeze fracture technique, illustrating the hexagonal arrangement of particles and pits. Magnification 127 000. (By courtesy of C. Stolinski).
B. Diagrammatic illustration of a gap junction in sectional view. In electron-micrographs the membranes of both cells appear as bilayers, and at gap junctions the space between the two neighbouring outer layers almost disappears.

could do so when in contact with a normal cell, either directly or (to a decreasing extent) through a chain of mutant cells by nucleotide transfer. It is now firmly established that passage of ions, fluorescent dyes and nucleotides all occur through gap junctions and, where cells are in communication, there must be a free flow of ions and small molecules, e.g. the cyclic nucleotides such as cAMP between them; the individuality and specific character of cells therefore depends upon specific macromolecules which cannot be transferred in this way. Several workers, notably Furshpan and Potter in 1968, Lowenstein in 1973, and Sheridan, who reviewed the subject in 1976, have suggested that gap junctions play an important part in development through controlling patterns of communication between cells, and therefore the flow of morphogens and the establishment of gradients (see chapter 12). At the least, this would allow movement of morphogen directly from cell to cell rather than by way of the intercellular spaces, with correspondingly less disturbance of the signal, but it is possible that gap junctions play a more direct role, by control of the timing and specificity (e.g. between one type of cell and another) of their formation. They can indeed be assembled very rapidly, within minutes or less, and may probably be broken down in a regulated way, but whether they are actually essential controls in the development of regional differences or simply nonspecific channels of communication remains to be seen.

The subject of the relevance of extracellular materials to development is also a subject of increasing interest, largely arising from work on the axial organs in the chick embryo in the 1960s and 70s, and particularly in connection with their possible connection with inductive interactions (see chapter 9), e.g. the studies by O'Hare and by Hay and Meier in 1974 on GAG (glycosaminoglycans). Collagen is also probably important, especially as a substrate for cell movement (see chapter 5), and problems concerning its origin and role in embryonic development were discussed by Hay in 1973. It is likely to be a significant constituent of the basal lamina— the ultrastructural membrane on which many epithelia rest and over which cells may move preferentially. Thus the observations of Trelstad, Hay and Revel in 1967 showed that the epiblast and endoblast of the chick blastoderm are underlain by discontinuous basal laminae, and that the primary mesenchyme cells move out laterally by extending filopodia and attaching to the basal laminae, as if they were feeling their way along it. Where there are gaps in the basal lamina, the mesenchyme cells make close contacts by tight and gap junctions to the epiblast and endoblast cells, and mesenchyme cells form similar connections with each other; presumably

86 NEURULATION AND THE DEVELOPMENT OF THE EMBRYONIC AXIS

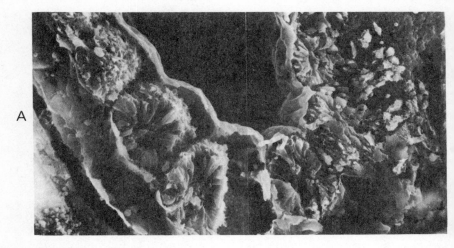

Figure 7.4 Somitogenesis in the mouse.
A. Scanning electron-micrograph of a sectioned 8-day mouse embryo showing development of the somites, the oldest of which are at the right (anterior end) and the most recently formed at the left (posterior end).

the close contacts serve as temporary anchorages, while the exploratory filopodia make more lateral contacts, then contract to draw the cell after them.

Avian neurulation and somitogenesis

It was established by Karfunkel in 1974 that the mechanism of neural tube formation, through elongation of neural plate cells and apical contraction, brought about by microtubules and microfilaments, is the same in bird as in amphibian embryos. More details have been added regarding the fusion of the neural folds in a scanning electron microscope study in 1975 by Bancroft and Bellairs, who showed that, as the folds approach each other in the dorsal mid-line, single threads are projected across from the surface of one cell to make contact with a similar thread, or the surface, of a cell from the other side. Establishing this degree of communication probably plays some part in guiding the folds into final apposition with each other, and final adjustments are made by tongue-like processes which insinuate themselves into gaps between the cells.

The first three pairs of somites are formed in the paraxial mesoderm beneath a broad region of the neural plate in the presumptive hind brain

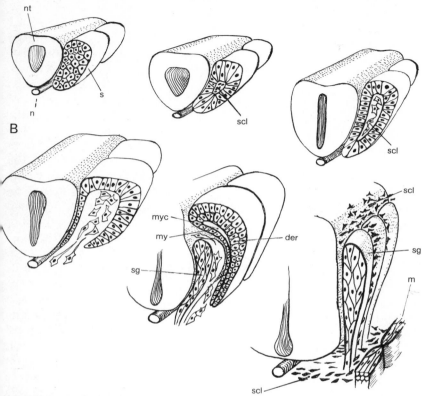

B. Diagrammatic sequence of events in somitogenesis. Mesenchyme cells in the vicinity of the neural tube (nt) form the somite (s)—an epithelial ball with a central cavity. Some of the products of cell division move into the cavity, giving the sclerotome (scl). Breakdown of the ventral and inner walls of the somites releases the sclerotome cells, which flow over neural tube and notochord (n). (Events up to this point can be seen in A). The remaining outer wall produces the dermatome (der), which gives the dermis of the skin, and the myotome (my), which gives the muscles (m), separated by the myocoele (myc). The spinal ganglion (sg), consisting of neurones of the sensory nerves of the dorsal root, arises opposite the dermatomes, and the sclerotome cells become arranged around them when they produce the chondrogenic condensations of the vertebrae.
Based upon O. P. Flind.

region, and more pairs are demarcated by intersomitic clefts as the node regresses, forming a series of cuboidal blocks, becoming triangular in cross section as they conform to their situation adjacent to the notochord and the neural tube on their inner face and to the overlying ectoderm to the outside. Older histological accounts of somite development have been supplemented by many more recent observations, including several of the general and ultrastructural studies on early chick development mentioned above, especially by Langman and Nelson in 1968, who, by following the synthesis of DNA in the developing somite, were able to clarify a number of problems concerned with the roles of mitosis and cell movement. A cavity (the sclerocoele) is established in the centre of the somite block and the cells become arranged as an epithelium around it, which has the appearance of being multilayered (pseudostratified) because the nuclei migrate from one end of the cell to the other in different phases of the division cycle. The process is essentially the same in mouse embryos, where it has been described in terms of the cellular interactions involved by Flint in 1977. The sclerocoele becomes filled by cells which are products of division within the somite epithelium, producing the loosely organized sclerotome. As development proceeds, the medial wall (against the neural tube and notochord) and ventral wall of the sclerotome become laminar instead of pseudostratified, and this transformation spreads up the walls between somites, leaving only the outer wall, against the dermis, as a thick pseudostratified epithelium. This forms an outer dermatome, and on its inner face its cells produce the myotome, separated from the dermatome by a space—the myocoele. The dermatome goes on to produce the dermal part of the skin, the epidermis of which is formed from the overlying ectoderm. The myotomes form the musculature of the back and limbs. Aggregates of nerve cells—the spinal ganglia—appear in the midst of the sclerotome cells opposite the dermatome; as the ganglia with their associated nerves expand, they cleave a path through the sclerotome, so that the sclerotome cells are compressed to form dense tissue between the succeeding ganglia and at the level of the intermyotome septa. This and other displacements of the sclerotome cells have the effect of moulding them roughly to the form of the cartilage rudiments of the vertebrae into which they will differentiate.

Cells of the neural crest

The appearance of neural-crest cells during neurulation in *Xenopus* was mentioned above, and such cells are produced in all vertebrate embryos from the marginal cells of the neural plate. They are carried with the neural folds until the latter fuse, when the neural crest, consisting of loosely organized cells, is formed along each side of the dorsal mid-line at the top of the neural tube. The subsequent development of these cells is one of the most remarkable stories in developmental biology and presents problems which still remain to be solved, for they now undertake a dispersal and migration to a wide variety of embryonic sites, some at very great distances away, and there differentiate into a comparable variety of distinct cell types. The broad facts were established in experimental studies by many early workers, including R. J. Harrison, and summarized by Hörstadius in 1950, but the modern phase of investigations came with the introduction of the use of better methods of cell labelling, notably in 1963 by Weston who used tritiated-thymidine-labelled grafts to trace the paths of migration and the fates of trunk neural-crest cells in the chick, and by Le Douarin and colleagues in studies using cephalic and trunk cells of the Japanese quail grafted into chick embryos, which she summarized in 1976. All these investigations confirm that the following at least are derived from the neural crest: the ganglia of the cranial sensory nerves and the dorsal root ganglia of the trunk; the parasympathetic ganglia innervating the gut and the ganglionic (chromaffin) cells of the adrenal medulla; much of the mesenchyme of the head (mesectoderm), including that from which the visceral skeleton (jaws and gills) arises; the pigment cells (melanocytes) of the epidermis, including those producing the pigmented patterns of hairs and feathers. The Schwann cells which produce the myelin sheath of medullated nerves have been thought to be crest cells, but they may have moved out from the neural tube.

Many problems remain partially or completely unsolved: e.g. how far the cells are determined when they leave the crest, and how far determination depends upon factors in the site they come to occupy; how far migration is at random and to what degree it is directed; and if it is not at random what leads to orientated movement over long distances?

Contact guidance has been proposed, e.g. over the ectodermal basal lamina for cells moving over the dermatome. Weston concluded that cells moved out from the crest along one of two routes: in the first cells move ventrally between the developing somite and the neural tube, and this stream divides into two at about the middle of the somite, one continuing in the same direction, while other cells migrate through the body of the somite—this route is now generally accepted; there is, however, some doubt about the second route, in which cells are supposed to move directly into the ectoderm after leaving the crest. Teillet and Le Douarin in 1970 using the quail/chick chimaera method, and Mayer in 1973 using mouse-mutant/wildtype chimaeras (see chapter 13), concluded that all neural-crest cell migration is by way of the mesoderm, and that melanoblasts move into the ectoderm at a later stage. The relation between patterns of skin pigmentation in amphibians and the patterns of cell behaviour which underlie them has been analyzed in a classic study by Twitty and Niu in

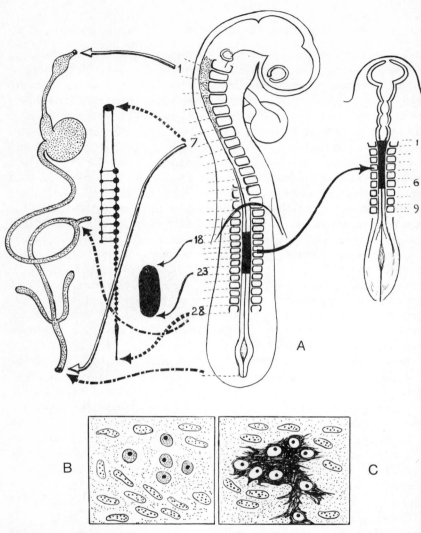

Figure 7.5 Differentiation of neural-crest cells as neurones of the autonomic nervous system or as adrenomedullary cells.
A. A quail embryo at the 28 somite stage. On the left are indicated the fates of some of the cells migrating out from the neural crest at various somite levels. Level 1-7 produces enteric ganglia of the preumbilical gut and innervation of the gut; from 7 posteriorly gives the orthosympathetic chain of ganglia; 28 posteriorly gives the ganglia of the postumbilical gut. The adreno-medullary cells are derived from level 18-23; if this part of the neural tube is transplanted to a chick embryo *(right)* at the 1-6 somite level, the neural-crest quail cells will appear in the wall of the gut, distinguished by their characteristic staining properties (B) and will differentiate as neurones of the gut plexuses, indicated by silver staining (C). Based upon N. Le Douarin.

1954, admirably summarized in Twitty's very delightful scientific autobiography of 1966.

The question of whether cells are determined to follow a particular developmental pathway when they leave the crest, or only after reaching their site of differentiation, has been clearly resolved by Le Douarin and colleagues in the case of those which become either neuroblasts of the parasympathetic ganglia innervating the intestine, or chromaffin cells of the adrenal medulla; these differ in respect of the neurotransmitter they synthesize, the first being cholinergic and the second adrenergic. Here the sites of origin in the neural crest for each type have been identified, and presumptive "adrenal" regions of quail neural tube grafted to the chick at the presumptive "parasympathetic" level; in this case these presumptive adrenergic cells enter the gut mesenchyme, differentiate as parasympathetic ganglion cells, and exhibit all the physiological properties of cholinergic neurones. In this case at least the decisive influence of the terminal tissue environment is established.

CHAPTER EIGHT

DETERMINATION AND DIFFERENTIATION

The molecular basis of cell differentiation

THE FULLY-DEVELOPED ORGANISM CONSISTS OF A NUMBER OF DIFFERENT sorts of cells, about 200 in vertebrates, produced in the process known as cyto-differentiation (or simply differentiation), which usually involves striking changes in cell shape, the appearance of new subcellular organelles or rearrangement of old ones, synthesis of specific products in the cytoplasm (e.g. haemoglobin in red blood cells), and in some cases their release through the cell membrane (e.g. extracellular matrix in cartilage). This process depends upon the implementation of particular genetic programmes, either by activation of some genes and inactivation of others at the level of transcription, by controlling production of the various messenger RNAs, or by control of their translation into enzyme and structural proteins, e.g. by modification or selection of mRNA, or by controlling ribosomal function in processing mRNA, or in releasing polypeptide chains. One of the most important objectives of current work in developmental biology is to explain differentiation in such molecular terms.

The epigenetic landscape: a model of determination and differentation

In his pioneering analysis of the relation of genetics to embryology in 1940, Waddington put forward a useful visual model of the developmental events leading to differentiation which he called the *epigenetic landscape*. It consists of a number of diverging pathways running downhill through a series of valleys, each of which terminates in an isolated location which represents a particular differentiated state. A ball, representing a cell's cytoplasmic state, place at the top of the landscape has the potentiality of ending up in any of the terminal locations but, as it rolls down and is deflected into one pathway or another, at each point of divergence its potential becomes more and more restricted until it enters a final pathway

DETERMINATION AND DIFFERENTIATION

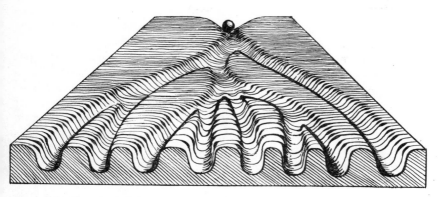

Figure 8.1 The epigenetic landscape model of determination and differentiation.
The genetic system controlling development provides a sequence of diverging valleys (pathways of determination) down which the cell rolls to its ultimate differentiated state. At first there are many pathways open to it, but the number becomes increasingly restricted as development proceeds.
Based upon C. H. Waddington.

leading to a unique differentiated state. This process of restriction of possible fates is known as *determination*, and the cell in its final pathway is said to be *determined*.

The topography of the landscape represents the total genetic programme of the organism; the slopes which lead the balls from their original positions into the valleys and keep them in the pathways at the bottom of them (*canalization* in Waddington's terminology) represent the various regulating mechanisms which produce normal development, in spite of disturbances which tend to force it off course. Which pathway is taken at any point of divergence depends upon cues supplied by the cell's microenvironment, e.g. inductive interaction with neighbouring cells or extracellular matrix, or response to concentration levels of a morphogen or a hormone. In many cases the choice is a binary one, e.g. in early mouse embryos blastomeres are determined as trophoblast cells or inner cell mass cells, suggesting simple genetic switch mechanisms, but there is no reason to suppose that in as many others the choice is not between multiple alternatives. The landscape will vary greatly, even in its general features, in different organims; in mosaic embryos entrance into the final pathways of determination will occur very early, through inclusion in the cells of the special plasm described in chapter 2, whereas in highly regulative embryos it will—except for the special case of the germ cells—be long delayed.

The relation between determination and differentiation poses fundamental problems, not always easy to define. It is important to notice in the first place that points of visible divergence in development do not necessarily correspond to points of divergence in the epigenetic landscape; at neurulation, ectoderm cells will look and behave differently according to whether they are going to become neural or epidermal, but this tells us nothing about whether they have just been switched into one or other of these pathways or have been in different determined states since early cleavage; to do that it is necessary to do an experiment, to see what the cells would have done in a different environment, either in the embryo or—what allows more sophisticated investigations—in cell or tissue culture.

In its fully differentiated state a cell is distinguished from cells of other types by characteristic histological features visible in the light microscope, expressing its particular phenotype. In this state of overt differentiation it is either functionally active or ready to function when required in postembryonic life. Some cells reach a state of differentiation when their histological type is recognizable but where the ultimate changes are not made until much later, as in the case of mammary gland epithelium which does not undergo its final differentiation until stimulated to do so by hormonal changes in pregnancy. In all cases, overt differentiation is preceded by a period in which changes are occurring but are only detectable using more elaborate methods of observation, and biochemical tests may make it possible to recognize cells as differentiated in this sense long before there is any visible appearance of it.

States of determination must themselves depend upon some, possibly extremely subtle, biochemical basis in the cytoplasm, and there is no general agreement as to whether a clear line can be drawn between a determined state and a differentiated state as it might be recognized by some all-seeing superbiochemist. Wolpert, whose work on pattern determination will be described later, has pointed out that cartilage as a tissue is identical in respect of cytodifferentiation in the various digits of the embryonic hand, but that in respect of its pattern of growth that of the thumb is in a different determined state from that of the little finger, only a part of which is expressed as cytodifferentiation. In such cases differentiation, in the ordinary sense, is clearly different from determination and for many purposes it is better to maintain the distinction.

In Waddington's model the final differentiated state lies at the end of a determination pathway, i.e differentiation follows determination. The question arises: can a cell in a particular pathway be rolled back up to the point of divergence and into another pathway, or even, can this occur in a cell which has arrived at overt differentiation? The problems arising from these questions have to do with the stability of the determined state. Answers can only be given for particular cases, obtained by testing the responses of cells to different environmental situations.

The cell at a point of divergence enters one pathway or another, depending upon some environmental cue. Its state of determination is labile and in one environment, in the embryo or in culture, it will take one way, in a different environment another; but tested after full commitment to one it will go on to differentiation as the type to which that pathway is leading, whatever the environment. For example, we shall see in chapter 11 that (at least according to many accounts) an early limb-bud mesenchyme cell may differentiate as cartilage in its original situation, but as muscle in a different one; but a cell from the same position in a later limb bud will differentiate as cartilage, no matter what site it is transplanted to, or in what medium it is cultured.

How stable is the determined state?

In certain situations some apparently differentiated cells may undergo a change of character which Weiss in 1939 termed modulation, e.g. Fell and Mellanby in 1953 showed that skin epithelial cells cultured in one medium, with excess vitamin A, become mucus-secreting, whereas with a deficiency they secrete keratin. But such changes are reversible, and the change is not into a different cell type but only into a different form of the same type; the term *dedifferentiation*, which was sometimes used in connection with this phenomenon, is misleading since there is no reversion to the point of divergence. In terms of the model it represents a rolling around at the terminal location; its determined state remains the same. In the case of the cultured skin, the environmental effect is upon the epithelial stem cells (see chapter 10) and is not observed directly in them but in the daughter cells they produce, which may be mucus- or keratin-secreting, but not both. The stem cell is determined as skin epithelium, but is still labile as regards the precise type of epithelial cell it may give rise to. A type of modulation frequently occurs in cell culture, in which a wide variety of cell types— muscle, liver, cartilage—may assume the form and behaviour of the fibroblasts described in chapter 5, and once this was taken as evidence that the determined state was a property only of cells in aggregates. In the 1960s the work of Konigsberg and others revealed that with improved culture techniques those cells retained their specific phenotypes, and consequently that whatever the basis of determination it is an autonomous cell character.

Metaplasia and transdifferentiation

However, there are some unusual cases, falling into two categories, in which the cell is rolled back to the point of divergence and over into a different pathway. One, termed *transdetermination* and discussed in chapter 15 (dealing with insect development), occurs when undifferentiated but determined imagined disc cells are maintained proliferating over many cell generations until put into a situation leading to differentiation, when a few differentiate as cells of a different disc type. The other, termed *metaplasia*, which is usually found in association with regeneration, occurs when fully differentiated cells undergo a true dedifferentiation process and redifferentiate as cells of a different sort. This phenomenon, which they term *transdifferentiation* in order to highlight comparisons between it and transdetermination, has been investigated by Eguchi and colleagues in 1973 and subsequently, using the techniques of clonal cell culture in which the fate of single cells and their progeny may be followed. The biological system they have used consists of tissues of the vertebrate eye, particularly the developing embryonic eye of avian embryos, and regeneration in the eye of adult urodele amphibians. In the latter, it has been known since 1895, when it became known as *Wolffian regeneration* after one of its discoverers, that if the lens is removed a new one may be formed from the neighbouring iris (but only dorsal iris) epithelium. In the organism, this phenomenon involves inductive interactions which complicate analysis of the differentiation process, but Eguchi, Abe and Watanabe have obtained differentiation of lentoids—structures shown to be lens-like by electron microscopy and immunological techniques—from dissociated newt iris cells in a simple cell culture system. Here, lentoids formed equally from cells of dorsal and ventral iris; this difference may be related to the fact that, in the operation on the organism, removing the lens loosens cell associations in the dorsal margin but not in the ventral margin of the iris, whereas in cultures all cells, from whichever source, are dispersed by enzymic dissociation; this suggests that changes at the cell surface provide significant signals in both cases. Other metaplastic changes have also been demonstrated in the eye, in adult urodeles and in early avian embryos. The Japanese workers have obtained almost all of these in cell cultures maintained for many cell generations; they have established that there is a sequence of transdifferentiation, from retinal pigment cells to neural retinal cells and vice versa, from either of these to lens epithelial cells and from the latter to lentoids.

Earlier studies by Cahn and Cahn in 1966 on clonal cultures of chick

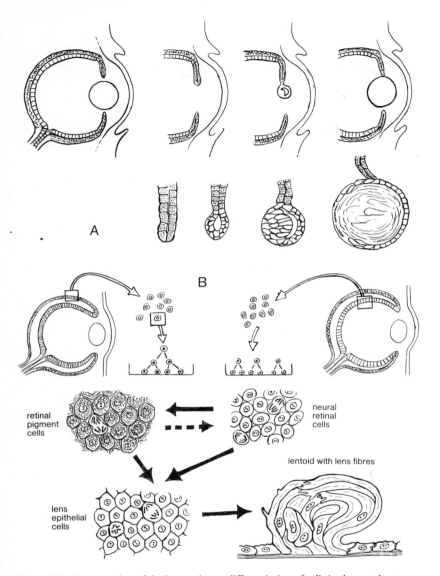

Figure 8.2 Regeneration of the lens and transdifferentiation of cells in the vertebrate eye.
A. Removal of the lens from an adult newt is followed by its regeneration from the neighbouring dorsal iris. Based upon H. Wachs.
B. Cultured retinal pigment cells *(left)* and neural retinal cells *(right)* from the chick embryo will transdifferentiate into lens epithelial cells, and further into lentoids. Retinal cells will also transdifferentiate into pigment cells, and probably (dotted arrow indicates this is not confirmed) vice versa. In the case of retinal pigment cells, clonal cultures derived from a single cell have been for the experiments.
Based upon T. S. Okada, Y. Itoh, K. Watanabe and G. Eguchi.

retinal pigment cells had shown that they retained their differentiated phenotype with complete stability over 50 cell generations, and this was accepted as reasonable evidence that these cells existed in an irreversibly determined state. The more recent work shows that such a state can never be demonstrated with complete certainty. The factor responsible for the switch to a different state after about the same period in Eguchi's cultures is not known, but certainly repeated replication of DNA is one of the prerequisites for transdifferentiation in culture.

Instability in these eye tissues appears to result from cell dispersal, and in some cases this leads to dedifferentiation followed by redifferentiation, either in a regeneration process or after prolonged cell culture, but still within a restricted range of cell types. Stability depends upon the establishment of particular patterns of cell associations. Determination here is clearly reversible; in terms of the model there is a point, or possibly several points, of divergence leading down gentle slopes to pathways which lead to the distinct differentiated states characteristic of the various eye tissues, but a number of unusual circumstances can push cells back and over into another pathway. They are irreversibly determined as eye cells, but reversibly determined as particular eye-cell types. There is sufficient stability to maintain the components of the eye in their proper states in normal circumstances, but enough flexibility to allow for repair after injury; natural selection in the course of evolution has produced a landscape which gives a degree of stability appropriate for this structure. The development of a cell culture system in which transdifferentiation can be observed and controlled with such precision is bound to throw light ultimately on the molecular basis of the determination process.

A relation between cell division and determination?

In both transdetermination and transdifferentiation there is a connection with cell proliferation, but it is also likely that determination in general may be related to cell division. As the cell moved down its pathway, it will of course be dividing, and sister cells may go down the same or different pathways. In many cases cytodifferentiation is incompatible with further division and mitosis ceases, but determination and partial differentiation are frequently accompanied by proliferation, as in the case of the various "blast" cells which generate more cells in a particular determined state which then undergo complete differentiation to give the fully differentiated "cytes", e.g. the melanoblasts and chondroblasts from which melanocytes (pigment cells) and chondrocytes (cartilage cells) are produced; myocardial cells, which eventually fuse to give the heart muscle fibres, actually commence beating as single and still mitosing cells. Holtzer has proposed that reprogramming of the cell occurs at the last mitosis preceding determination, and terms this a *quantal mitosis*, in contrast to the proliferative mitoses which lead simply to increased numbers of cells of the parent type. Holtzer has also suggested that cells produce two types of proteins; one, termed "house-keeping," consisting of those, such as those of the metabolic pathways, which are responsible for maintaining basic functions in most cells, and another termed "luxury" proteins which give differentiated cells their characteristic properties, e.g. myosin in muscle cells and haemoglobin in red blood cells. Treatment with 5-bromodeoxyuridine (BUdR) suppresses overt differentiation but allows and even enhances cell division, e.g. myoblasts continue to proliferate as individual spindle-shaped cells but do not go on to fuse to produce myotubes. The effect is reversible, which rules out a direct effect on the genome and implies an effect on some control mechanism. Since one may be affected but not the other, this suggests that the mechanism controlling "housekeeping" and "luxury" molecules may be different in type.

Erythropoiesis

In the generation of cell diversity, two separate but related problems arise. The first is the one dealt with particularly in this chapter, namely how a given cell of population of cells arrives at any point of divergence of its developmental fate with a potential for further development along some pathways but not others. The second is how the choice of particular pathways is imposed on individual cells in order to produce a coherently developing embryo; this is the problem of allocation of determined states and it will be dealt with in chapter 12 on pattern formation. But it is frequently difficult to disentangle the two in practice, since, as we saw in the analysis of transdifferentiation, the way in which cells are arranged in relation to each other in tissues or cultures plays an important part in their development. For this reason the study of determination and differentiation in blood cells is particularly important, since here there is virtually no spatial interaction of this kind, and the problems relating to them can be studied in isolation.

The modern phase of research into haemopoiesis—differentiation of the blood cells—began with work associated with the search for means of protection against the effects of whole body X-irradiation, leading to the discovery by McCulloch and Till in 1962 that the haemopoietic tissues could be repopulated by stem cells circulating in the blood. These cells (haemocytoblasts) arise from fixed mesenchymal cells and resemble the small amoeboid white blood cells known as lymphocytes. These stem cells are multipotent, producing most types of white blood cells (leucocytes) as well as the red (erythrocytes), but more is known about erythropoiesis. Differentiation occurs in several different sites, depending on the stage of development. In the mouse, Marks and Kovach showed in 1966 that erythropoiesis begins outside of the embryo, in the yolk sac; at 12 days the liver and spleen are the sites, and from 17 days chiefly the bone marrow, one site probably being populated by stem cells migrating from another. In adults, the bone marrow is the exclusive site, except in some anaemic stress conditions. In the course of differentiation the cells undergo a sequence of changes, first to the proerythroblast—which is the stage at which it becomes fully determined as a red-blood-cell precursor, then to basophilic erythroblast, then polychromatophilic erythroblasts—when small amounts of haemoglobin are synthesized but when the cells are still dividing, then reticulocytes (smaller, with much increased haemoglobin), then (in mammals by disintegration of the nucleus) into the erythrocyte itself.

The chief feature of differentiation in the red blood cell is, of course, the production of the protein haemoglobin, which is synthesized in such massive amounts that it was suspected that only gene amplification within the nucleus, i.e. production of copies of the relevant genes in enormous numbers, could account for it; but it has been shown that this is not the case, and that this extreme specialization of the erythrocyte must depend on control beyond the level of the gene. Several sorts of haemoglobin are produced in development, giving a sequence of embryonic, foetal and adult types. Thus in man, there are multiple types of HbE up to three months, HbF from then until shortly after birth, then two stable adult types, HbA and HbA_2. Allison showed in 1959 that all of the different human globins are products of different genes, so that there must be a complex molecular system controlling expression of these genes at particular times in normal development; and in the case of anaemic conditions the re-expression of some of them, e.g. for gamma globulin, characteristic of foetal haemoglobin, in adult humans. It is a popular but still unproved hypothesis that the number of divisions an erythrocyte cell has undergone determines which globin genes are expressed.

The chief difficulty in the way of studies on erythropoiesis has been its occurrence in a particularly inaccessible position, but this is now largely removed through the development of a technique by Farusawa, Ikawa and

DETERMINATION AND DIFFERENTIATION

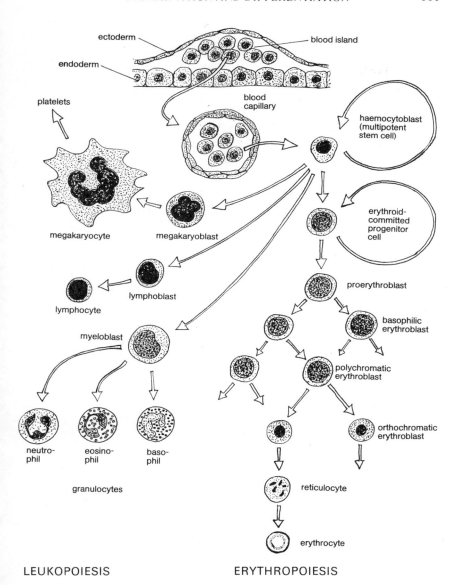

Figure 8.3 Erythropoiesis and leukopoiesis: the origin and differentiation of mammalian blood cells.
Partly based upon W. Bloom and D. W. Fawcett.

Sugano in 1971 for setting up cultures of erythroleukaemic cells, produced by infection of stem cells with a virus discovered by Friend in 1957, which exhibit some of the essential features of normal differentiation. One of the chief stages in erythrocyte differentiation from proerythroblast to erythroblast—is subject to control by the hormone erythropoietin. This hormone has almost no effect on the virus-infected cells, but the substance dimethylsulphoxide (DMSO) does induce many of the normal aspects of differentiation and, e.g. in the work of Paul and Hickey in 1974 who studied cell mutants differing in their reaction to it, this drug has become an important research tool in this system.

CHAPTER NINE
INDUCTIVE INTERACTIONS

Embryonic induction: evolution of a concept

TWO MAJOR DISCOVERIES AND THE IDEAS THAT STEMMED FROM THEM dominated the early years—well reviewed by Needham in 1950—of experimental embryology: the first was Driesch's discovery in 1891 that the sea urchin embryo was a "harmonious equipotential system", and the other Spemann's discovery of the phenomenon of induction, first in the development of the amphibian lens in 1901 and later, in 1924 with Mangold, in the case of what he called—since it initiated the development of the whole early embryonic axis—the *primary organizer*. The second led to such hopes of finding in it the key to all embryological mysteries that failure to isolate a specific chemical substance with all the directing and ordering properties that the organizer appeared to possess, and especially with Holtfreter's account in 1945 of how an enormous variety of entirely unspecific substances—organic acids, steroids, kaolin, methylene blue, sulphydryl compounds, which had nothing in common except the property of being toxic to subectodermal cells—produced neurulation in explants, that for a time a certain disillusionment with this approach to developmental biology set in.

This has passed, and induction is now seen in its proper perspective as a fundamental and highly—but not uniquely—important aspect of development, and as a term for a whole class of interactions between embryonic tissues, including a diversity of molecular mechanisms and systems of communication. Wherever two distinct groups of cells with different developmental histories come together in development, e.g. by rearrangement of the germ layers through morphogenetic movements, especially when both are organized as sheets (epithelia) or one is an epithelium and the other adjacent mesenchyme, e.g. between ectoderm and underlying dermal mesenchyme, or between one mesodermal rudiment and another, one is likely to affect the development of the other; the effect may be reciprocal. In a completely mosaic embryo, in which each region developed autonomously, this would not be so, but sufficiently detailed analysis would probably reveal inductions even in classic examples supposed to be of this type. For example, Raven in 1958 showed that the formation of the shell gland in the ectoderm of the snail *Lymnaea* is induced by contact with foregut endoderm. Whenever there is a capacity for regulation, i.e. in almost all embryos, much of development proceeds by this stimulus-response interaction between adjacent tissues, bringing them into line in time and space, and giving a tolerance of small variations and probably an evolutionary flexibility which a more mechanical system would preclude. The part played by inductions in determining the spatial patterns of development is rather limited (see chapter 12); as an approximation it can be said that an induction changes the overall development of a whole region of a cell sheet or cell mass, through the action of a signal or sequence of signals arising from a neighbouring tissue, but that the pattern of cellular differences arising in this region is determined by other signals, of positional information, which are generated within it.

Induction of the vertebrate lens

In vertebrates of all types the development of the eye is essentially the same. In chapter 7 we saw that neurulation produces an ectodermal neural tube with a wide lumen anteriorly in the presumptive brain region. Constrictions appear which divide the brain into fore-, mid- and hind sections, and from the forebrain a bulge (the optic vesicle) grows out from each side and comes into contact with the ectoderm at the surface of the head. The optic vesicle then becomes indented, forming the double-walled optic cup; subsequently the inner wall forms the optic retina and the outer wall, which becomes much thinner, the layer of retinal pigment cells (the only melanocytes which are not derived from the neural crest). The rim of the cup gives rise to the epithelial rudiment of the iris and the circular opening to the pupil of the eye. The stem of the cup is the optic stalk, and when the retinal cells differentiate as sensory neurones (retinal ganglion cells) their axons grow back along it to the brain, so that it is transformed into the optic nerve. The region of surface ectoderm in contact with the optic cup becomes thickened to form a placode, which sinks in to form a hollow vesicle which rounds off and becomes detached to give the lens rudiment. Later the inner vesicle wall becomes very much thicker than the outer; its cells elongate greatly, filling the lumen, then begin to synthesize crystallins—the characteristic luxury molecules of differentiated lens cells—to give the lens. Over the lens vesicle the ectoderm which remains at the surface becomes the outer layer of the transparent corneum, whose completion through immigration of neural-crest mesenchyme cells between the ectodermal epithelium and a thin mesodermal endothelium has been described in chapter 5.

Spemann showed that, in many amphibians, the lens is formed in the ectoderm only as a response to an inductive interaction resulting from contact with the optic cup. If the optic cup is removed before induction has occurred, the ectoderm in the eye region differentiates as epidermis, but if (within a certain period within which the ectoderm is said to be competent) presumptive flank ectoderm is grafted into this region, or if an optic cup is transplanted under the ectoderm in the flank region, this ectoderm, which would have become epidermis now produces a lens. More recent work by Jacobson in 1958 has shown that this classic induction process is in fact only the culmination of a sequence of inductive interactions (leading to the formation of the lens in the correct location at the right time), which begins with a reaction to the anterior archenteron roof in early gastrulation, followed by one to the presumptive heart mesoderm in the late gastrula, and concludes only with the response to contact with the optic cup. The influence of the three is similar and additive; the order of their activity may be changed experimentally, and one or other may be omitted so long as contact with the others is prolonged.

Instructive and permissive interactions

The optic cup/lens system illustrates several general points. Firstly, induction is followed by a phase of organogenesis, which may involve any one or several of the cellular activities we have discussed in connection with it (see chapter 5), succeeded by a phase of cytodifferentiation in which specific proteins are synthesized and the cells assume their characteristic differentiated form. Secondly, induction can be recognized only by experimental interference with the normal course of development, and it is the response of tissues to the presence or absence of other tissues which enables us to characterize the process. If the inducer is (1) present, the tissue it lies against will develop according to its presumptive fate but, if the inducer is (2) absent, it will not; but furthermore, (3) this inducer may be capable of producing the same development in cells whose presumptive fate is different—in this case, lens from presumptive flank epidermis. Where this third point can be shown experimentally, the induction is said to be *instructive* but, where only points 1 and 2 can be demonstrated, the induction may merely be *permissive*.

An instructive induction implies that the responding cells have been

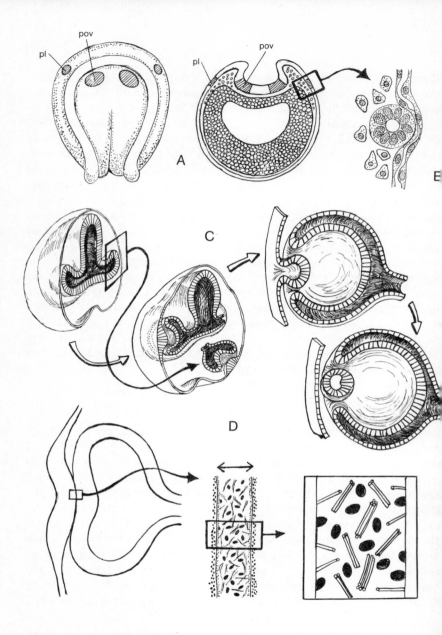

Figure 9.1 Lens induction in amphibian embryos.
A. Early neurula showing the presumptive optic vesicle regions (pov) in the neural plate and presumptive lens regions (pl) in the non-neural ectoderm, in dorsal view and in transverse section.

directed into a pathway of development which they would not otherwise have followed, i.e. that induction here has affected their state of determination; a permissive induction affects only the extent to which an already determined cell can express that state through differentiation in its particular pathway, and several permissive interactions will always succeed an instructive one.

The necessity for permissive interactions has especially complicated the interpretation of studies on induction done *in vitro*, where suboptimal conditions may inhibit differentiation which would take place if the culture environment were improved; thus Ellison and Lash showed in 1971 that chondrogenesis could occur in somitic mesenchyme in the absence of the neural inducer which had been thought, until then, to be required if the right culture conditions were provided. Again, cell proliferation is generally required for differentiation, and isolated epithelia in culture will not mitose; an underlying mesenchymal layer is required for this, but almost any mesenchyme will serve. In 1966 studies by Hauschka and Konigsberg stimulated a great interest in the effect of collagen in cell cultures; Dodson showed in 1967 that embryonic chick epithelium survived and keratinized on a collagen gel, and suggested that the epithelium is reacting to it in the same way as to the extracellular products of its normal dermis. Bernfield and colleagues followed up this lead in 1972 in the development of salivary gland epithelium, and found that not the collagen in the extracellular matrix but acid mucopolysaccharides in the basal lamina, whose synthesis depended in some specific way on the underlying mesenchyme, provided the crucial factor.

Besides the optic cup/lens system, many interactions between epidermis and underlying dermis (considered further in chapter 10) are clearly instructive, e.g. where chick dermis from feather-forming skin regions which is implanted beneath the corneal epithelium of the eye induces feather

---

B. Explanting presumptive lens ectoderm together with neighbouring portions of the archenteron roof and the lateral-plate mesoderm leads to induction of a small lens in culture. Based upon A. G. Jacobson.

C. The lens vesicle *(left)* left in place induces the formation of a lens in the presumptive lens ectoderm overlying it; where the lens vesicle has been removed *(right)*, no lens develops in the presumptive lens ectoderm; a lens develops wherever the extirpated optic vesicle is placed in any ectoderm which overlies it, providing that this ectoderm is in the "competent" stage.
Based upon H. Spemann.

D. A model of what may be occurring in the gap between the induced lens ectoderm and the inducing optic vesicle. Self-assembled ultrastructural filaments and globular aggregates within the matrix control passage of molecules between the cells.
Based upon R. W. Hendrix and J. Zwann.

formation in that tissue. But "instructive" in this sense does not necessarily, and almost certainly does not, involve a transference of a complex message; each cell at any point in its development will have only a restricted number of pathways open to it, perhaps even only two, and an instructive induction merely switches development into one or other of those pathways by activating a particular subroutine in the genetic programme. Thus though the response may be complex, the signal may be rather simple.

Examples of inductive interactions are found in development of the neural tube and somites (primary induction), which being more complicated will be considered last; in limb development (see chapter 11); in the development of the skin and its appendages (see chapter 10); and in the development of a wide variety of mesodermal tissues, e.g. in the differentiation of cartilage (chondrogenesis, see chapter 11), development of the pancreas, salivary and mammary glands, and of kidney tubules. Epitheliomesenchymal interactions such as the last four have been intensively studied (see the reports in the book edited by Fleischmajer and Billingham in 1968) on account of their suitability for *in vitro* culture and biochemical analysis, notably the pancreas by Rutter and Wessells in the 1960s, and Rutter and Ronzio in 1973. Most of the interactions between an epithelial and a mesenchymal component appear to be permissive in that, though a wide range of inducers may lead to the normal development of a particular structure, only one specific responding tissue will develop into a particular structure; the responding tissues appear to have been determined, but their differentiation may be nonspecifically stimulated by a variety of mesenchymes or, as in the case of kidney tubules, where neural tube will have the same effect, by other tissues. There is one striking exception, reported by Kratochwill in 1969, where mammary gland epithelium will branch in the manner of salivary gland epithelium if it is combined with salivary mesoderm, but even here its original biochemical properties persist, and in the appropriate hormonal situation it will secrete milk.

Induction of metanephric kidney tubules in vertebrates

The kidneys are formed in a strip of intermediate mesoderm on each side of the embryo which lies between the somites and the lateral plate. From this a series of tubules is formed, divided into three groups produced in sequence the pronephros, mesonephros and—in birds and mammals, in which it is the only one which survives into postembryonic life—the metanephros Tubules from the pro- and mesonephros open directly into a longitudinal Wolffian duct, but those of the metanephros open into a ureter which grows

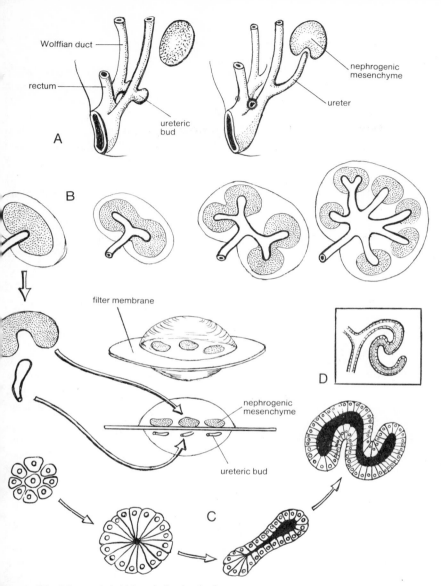

Figure 9.2 Metanephric kidney induction in the mouse.
A. Outgrowth of the ureteric bud anteriorly into the nephrogenic mesenchyme.
B. (left to right). Induction of tubule formation in the mesenchyme, beginning with condensation of the mesenchyme cells around the branching tips of the ureter rudiment.
C. Kidney tubulogenesis observed in nephrogenic mesenchyme cultured on a filter membrane, in proximity to a ureteric bud placed on the other side of the membrane.
D. Final stage of tubulogenesis, in which the tubule (stippled) becomes confluent with the ureteric branch which has induced its development.
Based upon L. Saxén et al.

forward from the posterior end of the Wolffian duct into the nephrogenic mesenchyme and there induces the mesenchyme to form the tubules. In a number of mouse and chick mutants, e.g. the *wingless* mutant of the fowl (see chapter 13), the ureteric bud fails to grow forward, so that it does not reach the mesenchyme, which therefore produces no tubules. This system has been the subject of detailed *in vitro* studies by Grobstein and Saxén and their colleagues. Formation of tubules has been obtained in transplants of the mesenchymal rudiment without the ureter to the anterior chamber of the eye, indicating that determination has occurred already and that the interaction is permissive, rather than instructive, but not so far in any *in vitro* conditions.

The cellular activity involved in the phase of organogenesis following induction has been investigated in detail by Saxén's group, reported in 1968. The inducing tissue, the ureter rudiment, grows into the loosely organized mesenchymal rudiment and forms a branching tree-like structure within it; the mesenchyme then condenses around the tips of the branches, forming dense aggregates within which the cells become elongated around a central slit. The aggregate then bends into an S-shape, and the slit becomes continuous with the interior of the ureter branch forming the completed tubule. In the first stages, the mesenchymal cells move at random, then become trapped within the condensations, within which they gradually lose their mobility. The condensations become first spherical, with radially orientated wedge-shaped cells, which suggests that one of the first results of induction is increased cell-to-cell adhesion. That there is some cell surface change is further supported by the acquisition, during the aggregation phase, of a resistance against polyoma-virus infection. More details have been added by the studies of Gossens and Unsworth in 1972.

Primary embryonic induction

In amphibians, determination of the neural ectoderm occurs between late gastrulation and early neurulation; if presumptive neural ectoderm from a gastrula is implanted into presumptive ventral ectoderm of another gastrula at the time of dorsal-lip invagination, it develops into skin in the same way as the surrounding epidermis; but if the same operation is performed on an early neurula, a neural plate is formed in that position as well as the normal one. Grafts of dorsal-lip material and the chorda mesoderm invaginated there will induce a new neural plate wherever they are implanted beneath competent ectoderm; a piece of the dorsal lip in-

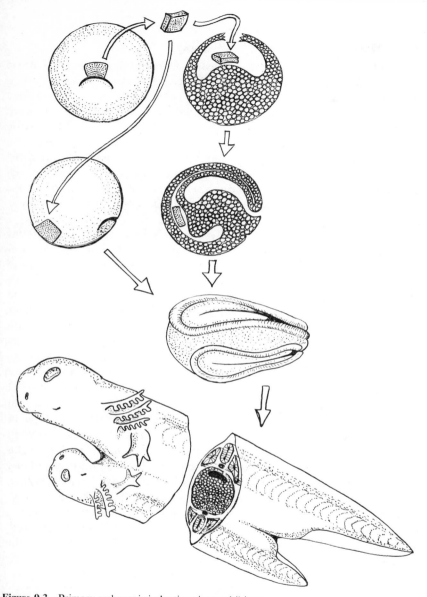

Figure 9.3 Primary embryonic inductions in amphibians.
A piece of dorsal blastopore lip material, taken from one newt early gastrula and implanted into the ventral region of another, causes the overlying host tissue to develop a second embryonic axis, producing a secondary embryo attached to the original one. An alternative method is to insert the blastopore material into the blastocoele of a late blastula, where the gastrulation movements will carry it into a ventral position adjacent to the host tissues. Based upon H. Spemann.

troduced into ventral ectoderm will form a secondary blastopore and move in as it would do at its original site, or it can be simply pushed into the blastocoele, when the morphogenetic movements will force it up against the ectoderm. All of the events of neurulation and somitogenesis follow, and then organogenesis, to give a complete secondary embryo. Most work done since the beginning of the 1950s, e.g. by Nieuwkoop and colleagues, and by Toivonen and Saxén, has been done not on whole embryos but using culture techniques in which competent ectoderm is wrapped around the inducer (or combination of inducers) and—since they can be obtained in large quantities—often using as artificial inducers material from quite different sources.

Neurulation does not produce a simple tube, but a tube with regional differences, and this appears to be produced by corresponding regional differences in inductive activity; pieces of chordamesoderm taken from the anterior end of the archenteron roof will produce secondary embryos with good head but poor tail development, and vice versa for pieces from the posterior end. Unnatural inducers will mimic these effects: guinea-pig liver cells or heat-inactivated Hela cells will produce neurulation leading to forebrain structures, and guinea-pig bone marrow will produce mesodermalization, i.e. it will cause ectodermal cells to produce mosodermal structures such as somites and notochord. A mixture of the two will induce hindbrain structures and spinal cord. This led Saxén and Toivonen in 1961 to propose a double-gradient model for primary induction, suggesting that two such factors exist in the chordamesoderm, and that the type of structure induced in a particular region depends upon the balance between the two at that level. The two factors probably act sequentially, neurulization occurring first, uniformly, by a factor produced by the invaginating dorsal-lip cells, which is then modified in a regional way by a mesodermal factor produced by the invaginated mesoderm of the archenteron roof. Where gradients are observed in inductive interactions, the distinction between induction and positional information models becomes blurred, and may disappear with more precise knowledge of the process at the molecular level.

The nature of the inductive signal

In all inductions the interacting tissues are in close proximity, but it is not easy to establish what the precise degree of contact is in particular cases. Grobstein, developing the *in vitro* metanephric mesenchyme/ureter (or neural tube) system in the 1960s, showed that induction could occur across a Millipore filter at least 30 μm thick and with pores of 0.45 μm diameter, in which no cytoplasmic process could be detected, suggesting that some transmissible factor was involved; but work by Wartiovaara and others in Saxén's laboratory showed that if Nucleopore filters (which have straight cylindrical pores of more uniform size) are used, long cytoplasmic processes can be seen extruding into them, suggesting that actual contact between the plasma membranes of inducer and responding cells can be made, and that previous transfilter results should be reinterpreted. Weiss suggested "contact orientation" between molecules of adjacent cells as a mechanism of induction, but proposed an idea that was further developed by Grobstein—that the two surfaces would not require to be in direct apposition, but that this type of interaction could also occur if they were separated by a narrow space filled with intercellular matrix secreted by the cells and containing a population of molecules whose free mobility was restrained. Investigating the nature of this extracellular matrix and its role in developmental interactions in a diversity of systems, is now an extremely active research field. Thus in the optic cup/lens system in the chick embryo, McKeehan established in the 1950s that the two cell layers became extremely difficult to separate in the region of contact, but that introducing agar blocks 20 μm thick between them did not prevent induction. Hendrix and Zwann showed, using the electron microscope and autoradiographic labelling in 1975, that there is indeed a gap, filled with material secreted by the cells during the period of lens induction—glycoproteins, glycosaminoglycans and collagen—which produce a characteristic matrix structure with globular aggregates and thin filamentous structures, probably produced by a process of molecular self-assembly. This structure may act in induction by controlling the passage of molecules between the cells and also—in the phase of organogenesis—by increasing adhesion between the lens placode and the optic cup, preventing the lateral expansion of the lens cell population and causing by simple cell crowding many of the characteristic cell shape changes which produce the lens vesicle.

Since work in the 1930s showed that in amphibians pieces of heat-killed mesoderm induced neurulation, and therefore that some chemical agent was involved, attempts to isolate and characterize diffusible factors of induction have never ceased. One possibility is that large information-carrying macromolecules might be transmitted from cell to cell; in the 1940s Brachet produced some evidence which suggested that RNA-rich particles were so transmitted in neural induction, and proposed that the process was comparable to virus infection. Proteins have also been proposed as agents; Yamada and Takata in 1961 extracted a mesoderm-

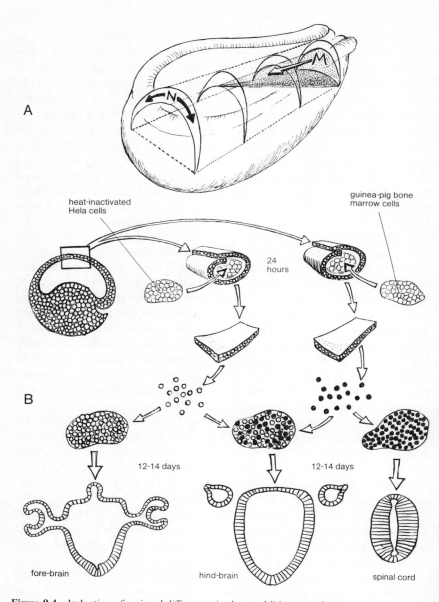

Figure 9.4 Induction of regional differences in the amphibian neural tube.
A. Saxén and Toivonen's double-gradient model of neural induction, superimposed on an amphibian neurula. A neurulizing factor, produced in the midline and spreading laterally, interacts with a mesodermal factor produced posteriorly and diffusing forward.

inducing protein fraction from the guinea-pig bone marrow used as an artificial inducer in the experiments described above, and in 1968 Tiedemann isolated a protein with similar properties from 9-day chick embryos. But all of these results are open to alternative interpretations and, in the light of current concepts of determination, and of the nature of cell communication in early embryos, it is considered more likely that no very complex informational transfer takes place. There is much interest in the investigations by Barth and Barth, e.g. in 1972, which indicate that, on the contrary, simple organic cations (Na^+, Ca^{2+} and Li^+) induce differentiation of neurones from presumptive ectoderm in anurans, and at a later stage of ectodermal competence the same ions induce development of these neurones into pigment cells. These ions are clearly acting by operating switch mechanisms within the responding cells, and it may well be that they function in this way in normal development, either directly or indirectly. In 1974 McMahon proposed a hypothesis for induction which incorporates this concept and links inducers with other chemical messengers, both in the embryo and the adult. This hypothesis suggests that molecules which function as neurotransmitters (e.g. serotonin, acetylcholine) in physiological activity, are secreted by the inducer cells and act at the surface of responding cells by controlling their intracellular concentration of cyclonucleotides—cyclic AMP (adenosine monophosphate) and cyclic GMP (guanosine monophosphate) (the so-called second-messengers of nonsteroid hormones) whose effects and metabolism are intricately interwoven with the ionic conditions within and outside of the cell, which will act upon the gene regulatory mechanism to lead development into particular developmental pathways. The role of neurotransmitters as information carriers in eukaryotes may therefore be much wider than previously supposed, beginning in evolution as intracellular messengers, then serving as intercellular messengers for the relatively slow communication of developmental information, and finally evolving as the agents of rapid intercellular neural communication in adults.

B. Some experimental evidence on which the model is based, using heat-inactivated Hela cells as neural factor, and guinea-pig bone marrow cells as mesodermal factor, enclosing them in rolls of presumptive ectoderm from an early gastrula. The first induces fore-brain structures, the second, spinal cord. If the induced ectoderm cells are disaggregated and then reaggregated, the same type of differentiation occurs, but if the two are mixed before reaggregation then hind-brain structures are produced, indicating that induction of the central nervous system is a 2-step process: first, ectodermal cells are determined towards either neurulized or mesodermalized cells, then these two cell types interact to determine the regional character of the CNS.
Based upon L. Saxén, S. Toivonen and T. Vainio.

CHAPTER TEN

DEVELOPMENT OF THE SKIN AND ITS APPENDAGES

Ectodermal and mesodermal components of the skin

THE VERTEBRATE INTEGUMENT IS MADE UP OF AN OUTER EPIDERMIS, which arises from the ectoderm, and an inner mesodermal dermis which becomes separated as a distinct layer from the underlying loose mesenchyme fairly late in development. The two together form the skin, the organ of the body which serves to cover and protect it, and from which more specialized structures, notably glands and hairs, feathers or scales, are produced.

The epidermis is made up almost entirely of keratocytes (or keratinocytes), with some melanocytes, the pigment cells derived from the neural crest. It originates as a single ectodermal layer, resting upon a basal lamina which becomes divided into basal and periderm layers, of which the periderm is lost early in embryogenesis. Thereafter the basal cells form the germinal layer, a single layer consisting largely of stem cells which by mitosis produce lineages of cells which build up further layers of cells with the oldest layer at the surface. Differentiation of the epidermal cells has been studied most in mammals, and chiefly in postembyronic stages, in which the stem cells of the germinal layer continue to divide throughout life, replacing the external cells which die and are sloughed off as they become heavily keratinized.

Keratocyte differentiation in mammals

Synthesis of the protein keratin is the chief feature of epidermal cell differentiation, in the course of which the cells change their columnar shape, becoming first polygonal, then flattened, and finally mere dead flakes of keratin. Differentiation begins when cells in the germinal layer lose contact with the basal lamina and enter the stratum spinosum, so called because cells here appear spiny owing to the presence of many microfilopodia which make contact and at their tips, where desmosomal junctions are established, with similar cytoplasmic extensions from neighbouring cells. In the cytoplasm there begin to appear ultrafine (about 7 nm thick) filaments, possibly derived from slightly thicker tonofilaments found in the basal cells, which will give the fibrous component of keratin. Some mitosis continues in this layer, but not beyond it. These two layers are often called jointly the Malpighian layer. The next one, the granular layer is characterized by the presence of heavily-staining keratohyalin granules—which are probably involved in keratin-matrix synthesis, but these are absent from the next layer, the stratum lucidum. Finally in the embryo is the stratum corneum, consisting of extremely flattened cells which lose their nuclei and die, and by combination of the fibrous and matrix components become transformed into flat plates of keratin; the desmosomes still persist, so these plates remain linked with their neighbours to form a continuous layer. In postembryonic stages there is still another layer, the stratum disjunctivum, in which the desmosomes break down and the keratin flakes fall away.

Though keratin synthesis is the distinguishing feature of most of its cells, the epidermis does also form mucus-secreting glands, and the capacity of epidermal stem cells for modulation between production of cells with these two types of activity, shown by Fell and Mellanby in 1953, has been mentioned in chapter 8.

Studying frozen sections of unfixed skin in 1969 and 1970, McKenzie discovered a striking architectural organization of the epidermal cells not revealed by conventional histological methods. He showed that the cells of the stratum corneum are stacked with great precision in columns, their edges overlapping in an intricately alternating way with those of neighbouring columns. From the surface, the columns present a honeycomb appearance. Stacking begins in the Malpighian layer, where the cells, being polygonal in shape, are small compared to the cross-sectional area of the column of flattened cells, so that each column is situated above about 10 Malpighian cells. It appears that the most actively mitosing cells are at the periphery of each group, and that cells about to undergo differentiation move centripetally into the centre of the group, then outwards along the axis of the column, flattening and contributing to it as differentiation occurs. To generate this very precise arrangement, production of cells and their movements must be very accurately controlled and Christophers showed in 1972, in the guinea-pig, that if the proliferation rate in the basal layer is increased by rubbing the skin this ordered stacking disappears.

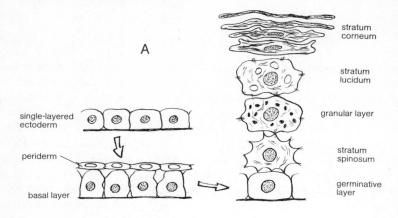

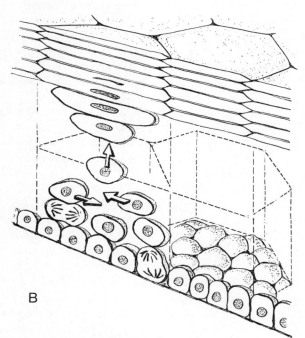

Figure 10.1 Development of the epidermis in mammals.
A. Diagram to show development of the mammalian epidermis from the single-layered ectoderm, through a two-layered stage in early development, to the multilayered structure of the end of embryonic development. Each layer in this final phase except for the germinative layer is several cells deep in reality.
B. Keratocyte differentiation, illustrating the movement of cells derived from the germinative layer to produce highly ordered polygonal columns of flattened keratinized cells in the stratum corneum.
Based upon E. Christophers.

Control of epidermal proliferation

In adults, the epidermis exists in a state of dynamic equilibrium, continually losing cells at its surface and continually replacing them by mitosis in cells of the basal layer. Maintenance of this steady state depends upon a balance existing between cells that remain basal and proliferative, and cells that move out and differentiate. The kinetics of this process has been studied by a number of workers, e.g. by Iversen, Bjerknes and Devik in 1968. By labelling epidermal cells in a strain of hairless mice with tritiated thymidine, they showed that, when a basal cell divides, both daughter cells, or one, or neither may become committed to differentiation; if a cell does become so committed, it remains in the basal layer for another 60 hours before moving out and takes a further 84 hours before arriving at the outer surface. About 40% of cells will move out in this way without undergoing another division, the remaining 60%, consisting of cells which will divide at least once more, constitute the progenitor cell population. This population of stem cells is maintained, in a determined but undifferentiated state, throughout the organism's life.

In 1957 Weiss and Kavanau proposed a model for growth control based upon the concept of negative feed-back, i.e. that a factor produced by a mass of differentiated cells, and therefore increasing in proportion to the number of cells, would inhibit proliferation in the cells which generated them. In 1962 Bullough obtained evidence that just such a mechanism acted in the control of epidermal cell proliferation: a factor, called a *chalone*, is secreted by the differentiating keratocytes, which inhibits proliferation in the basal layer in proportion to the amount produced. All tissues produce a chalone specific to themselves, which has no effect upon other tissues; but chalones are not species- or even class-specific, and extracts from a wide variety of vertebrates, including the cod, will inhibit proliferation in the epidermis of the mouse, though an extract from the mouse dermis will not. The implications of chalone theory are important, particularly, as Bullough has proposed, e.g. with Deol in 1971, for its possible applications in tumour-therapy, but there are still large areas of uncertainty in this field.

Many other factors besides chalones affect proliferation and differentiation of keratocytes, e.g. Vitamin A increases proliferation and hydrocortisone depresses it. But the factor with the most dramatic effect is the so-called epidermal growth factor (EGF) whose effects were described by Cohen and Elliott in 1963. EGF is a protein contained in extracts of the submaxillary gland of the male rat which has a strong stimulating effect upon epidermal growth and keratinization, causing, when injected into newly born mice, precocious eruption of the incisor teeth and opening of the eyelids after 6 days rather then the normal 12 or 13. This factor again is not species-specific.

Epidermal-mesodermal interactions in skin development

The dermis consists of a rather densely packed mesenchyme with an intercellular ground substance which contains many collagen fibres, formed by a process of molecular self-assembly from simple tropocollagen units produced by the fibroblast, i.e. the dermal mesenchyme, cells. Self-assembly of this type, i.e. the process whereby elementary protein units become aggregated and arranged to form more complex proteins without interaction with any other complex system, and with no appeal to any mechanism beyond basic physical forces and principles such as the minimization of free energy, constitutes a fundamental activity in development, upon which those processes which are beyond (temporarily some would say) this type of explanation are built up. The self-assembly of collagen, described by Gross in 1956 remains an important system for detailed physical and chemical analysis of this process.

Since almost nothing has been said about the dermis, it might be thought to have little but a supportive role in the integument, but this is not the case. It is, on the contrary, the mesoderm which determines the regional character of the skin, as shown in the transplantation experiments which Billingham and Silvers made in the adult guinea-pig in 1967. Exchanges were made between epidermal and mesodermal components of the trunk, where the epidermis is thin, the ear, where it is twice as thick, and the sole of the foot, where the stratum corneum is exceedingly thick; in each case the epidermis took on the regional character of the dermis.

Cutaneous appendages: hairs, feathers and scales

The skin appendages—hairs in mammals, feathers and scales (on the legs) in birds, and scales in reptiles—develop in essentially the same way, by local growth and activity in both epidermis and mesoderm. In the case of hairs, a deep pit (the follicle) is produced by epidermal ingrowth into the dermis, in which the hair is produced; in feathers growth is chiefly outward, to form a papilla. Here we shall consider chiefly feather development.

The feather of adult birds is a structure of great beauty and complexity,

DEVELOPMENT OF THE SKIN AND ITS APPENDAGES

consisting, over most of the body, of a central shaft or rachis with a vane on each side of it. Each vane is made up of parallel barbs, arising from the shaft at an angle, from which in turn arise parallel barbules anteriorly and prosteriorly. Each barbule consists of a dorsal ridge and a ventral membrane, the membrane of the anterior barbule forming a number of hooks, while the posterior barbule ends in a single saw-edged curve. The hooks of each anterior barbule overlap and grapple with the dorsal ridges of several posterior barbules, whose saw-edged tips prevent the hooks slipping out of position until they are forced. This arrangement produces a zip-fastener type of mechanism, producing a structure of great strength and flexibility, exactly adapted to its various functions. An analysis of the morphogenesis of the fine details of feather structure, which must involve extraordinarily precisely controlled events at the cellular level, has never been undertaken, but the origin of rachis and barbs from a proliferating collar of epidermal cells has been the subject of many studies, e.g. those of Cohen and 'Espinasse in 1961. The entire feather, and the hair in mammals, is formed from keratinized epidermal cells, and the mesodermal component persists only as a small basal dermal papilla which becomes active again when the feather or hair is lost and must be renewed, as happens irregularly throughout adult life, or regularly at a seasonal moult. But again it is the dermis which chiefly controls what will be formed from the epidermis, which evidently has a repertoire of structures it is capable of producing, with the dermis selecting which one is actually to appear. Thus, in 1965 Rawles showed that even the epidermis of the beak will produce feathers with appropriate dermis and, more surprisingly still, Coulombre and Coulombre in 1971 showed that the corneal epithelium of the eye in the chick would produce feathers, indicating a type of metaplastic transformation. In these studies it was shown that mouse dermis will induce the earliest stages of feather development in chick corneal epithelium, demonstrating that the epidermis can respond to a signal from dermis of an organism from a different class, but only by development within its own species' repertoire. Dhouailly and Sengel have shown in an extensive series of dermal/epidermal exchanges between species (chick and duck) and between classes (mammal, bird and reptile) that the initial dermal induction which is non-specific must be succeeded by a second dermal induction which is class-specific in order to produce a fully formed hair, feather or scale. These workers have also shown that, when chick/duck wing-dermis/epidermis combinations are made, all of the structural characteristics and even the distribution pattern of the different types of wing feather conforms to that of the dermal-donor species, with the

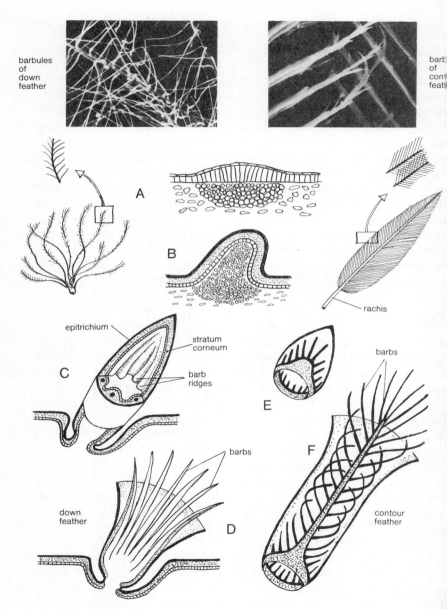

Figure 10.2 The development of feathers in the fowl.
Feathers of chicks are mostly down feathers *(left)* with barbs arising together from a short base; in adults they are mostly contour feathers *(right)*, whose barbs arise from an elongate rachis. On either side of each barb is a row of minute barbules which are provided in the case of

exception of the minute structure of the barbules which show all the characteristics of the epidermal donor.

The capacity described in chapter 5 of dissociated cells to reaggregate and rearrange themselves to reconstitute embryonic structures was demonstrated by Garber and Moscona in 1964 to occur to a remarkable extent in dissociated chick and mouse skin when the cells, or reaggregates made from the cells, were allowed to develop on the chorioallantoic membrane of the chick; aggregates of chick skin cells formed well-developed feathers, while aggregates of mouse skin cells formed thickly stratified layers of keratinized skin, hair follicles and rudiments of sebaceous glands. Mixed suspensions of cells produced a chimaeric skin with chick and mouse cells mixed singly or in groups, together with hairs and sebaceous glands, but never any feathers which, for an as yet unknown reason, are entirely suppressed by the presence of even a small proportion of mouse skin cells or, as shown by Moscona and Moscona in 1965, any other tissue cells from the mouse or the chick.

The development of patterns of feather arrangement

The dissociation and reconstitution of developing feathers is also shown in studies by Novel in 1973, in which pieces of skin from the back of the chick embryo were taken, the epidermis separated from the dermis after mild treatment with trypsin, and then allowed to develop in culture with the epidermis replaced on the dermis but rotated through 90° or 180°. Under these conditions the first rudiments of the feathers, visible at the time of

---

the contour feathers with elaborate hooks on the anterior barbules which grip corresponding ridges on the posterior barbules of the barb in front, forming an effective zip-fastener mechanism *(top right)*. There are no hooks or ridges on the barbules of a down feather, producing its characteristic fluffiness *(top left)*.
In both types development is similar in the early stages:
A. Formation of a thickened epidermal placode and an underlying mesenchymal condensation in the dermis.
B. A projecting feather papilla is formed, with posteriorly directed orientation. The epidermal component consists of the germinative layer, the stratum corneum and an outer epitrichium. Further development is simplest in the down feather.
C. The epidermal component becomes ridged and the interior of the ridges becomes heavily keratinized to give the barbs.
D. The epitrichium breaks and the barbs splay out.
E, F. Development of the contour feather is essentially similar, but as the feather papilla enlarges the growth of the epidermis produces a long outward extension of the epidermal collar at its base. The dermis forms a persistent nutritive core at the base, supporting the growth of new feathers when the old ones are moulted; the externally visible part of the feather is purely ectodermal.

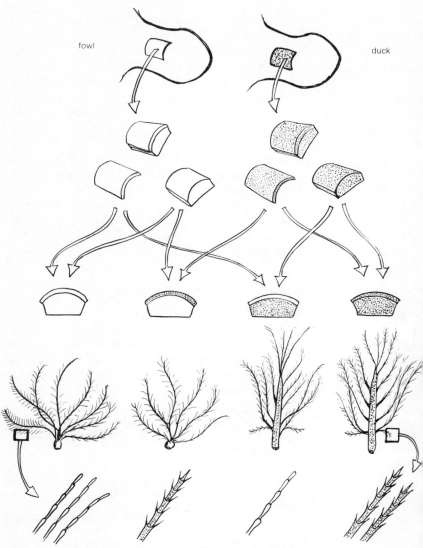

Figure 10.3 Interspecific ectodermal-mesodermal interactions in feather development. Pieces of skin are taken from limb buds of fowl *(left)* and duck *(right)* embryos and the ectodermal and mesodermal components separated, then recombined, as they were and reciprocally, and cultured. The type of down feather which is produced is always determined by the dermal component (in ducks the down feather has a rachis) except in respect of the minute structure of the barbules, which is determined by the epidermal component.
Based upon P. Sengel.

explanation, disappeared; they reappeared later, but arranged in a different pattern. This experiment throws considerable light on how their overall pattern is established.

Early morphogenetic events in feather development have been described in detail by Wessells in 1965, but here it is sufficient to say that the first visible sign is a slight thickening of the ectoderm, producing a disc-shaped placode, followed by the appearance of a dermal condensation of mesoderm cells beneath it. Wessells believed that this condensation could be accounted for solely by a very slight increment of cell division in this region, but others (and Ede, Hinchliffe and Mees produced some evidence for it in 1971) believe that there is in addition an actual aggregation by cell movement into these condensations.

In the skin of the back of a chick embryo the first feather rudiments appear equally spaced in a single row along the mid-dorsal line. Shortly afterwards, another row appears laterally on each side of the first, then another row on each side, until the whole back region is covered with feather rudiments. As each row appears, the rudiments are equally spaced along it, but alternating with those of the previous row, so that a very regular pattern is built up. In Novel's experiments the square of skin taken for explantation included the mid-dorsal row along one side, and extended laterally towards the flank. Depending on the age of the embryo, a single row of rudiments, or two or more, had appeared at the time the skin was removed to culture. The reorganization of rudiments after a few days in culture showed that the epidermis had no effect upon the arrangement of the new rudiments except in one particular—that the orientation of individual feather rudiments, which normally point backwards, followed the orientation of the epidermis, so that if that had been rotated through 90° the rudiments grew out pointing sideways, and if the rotations had been through 180° they pointed forwards; this polarity is therefore determined in the epidermis. The overall arrangement of the reorganized rudiments, however, was determined solely by the dermis but, except in the case where only one row had been present, the pattern was not simply a reconstitution of the old one; instead, the first new row of rudiments to appear did so in the position of the youngest row of the old rudiments, and further rows appeared in sequence on each side of that. Sengel in 1975 concluded that the arrangement of feather rudiments is not predetermined in unpatterned skin, but becomes gradually organized as the rudiments of successive feather rows are formed. Linsenmayer in 1972 also concluded from his own work that the control of feather pattern formation depends only on short-range interactions between developing rudiments, but in 1972 Harris

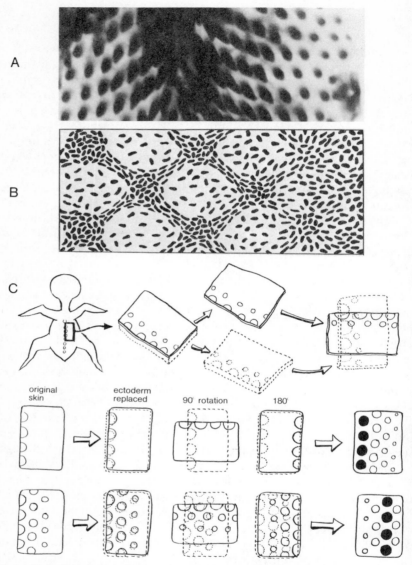

Figure 10.4 The pattern of feather arrangement in the developing fowl embryo.
A. Dorsal skin from an 8-day embryo, showing sequence and arrangement of feather papillae symmetrically about the mid-line.
B. Appearance of dermal cells in skin from an 8-day embryo. The mid-line of the embryo is at the left, and the dermal feather condensations are established most clearly here, with tracts of orientated cells between the condensations. Laterally, to the right, condensations and tracts are only beginning to appear.
C. An experiment to show the role of the dermis in determining feather pattern. Explanation in text.
Based upon G. Novel.

advanced very interesting arguments against this, based upon crystallographic analysis of dislocations which occur in the pattern.

The origin of the precise arrangement of these rudiments, each at the angle of a lozenge-shaped quadrilateral, is still an open question. In mammals the arrangement of hairs is much less obviously ordered, though it is not random. Based on his studies on the arrangement of hair follicles in sheep, Claxton produced a model in which differentiation of a rudiment at any site, initiated by some chance event in a single cell, suppresses further initiation in a defined area around it, which gives—what is found in mammals—regularity in respect of spacing but no constant pattern. But introduction of some further simple constraints can produce, as shown by Ede in 1972, the feather pattern from this type of model. A different suggestion was made in 1967 by Stuart and Moscona, who showed the existence of a collagen lattice in the dermis, connecting the rudiments and along which the mesenchymal cells might be led through contact-guided movement to form the dermal condensations. It is not, however, quite certain whether the collagen lattice appears before the dermal condensations appear, or is secreted by mesenchyme cells which have become orientated in relation to the condensations. Sengel, whose work has provided much of the information we have in this field, believes that the pattern of dorsal feathers arises from the close-packing of the rudiments in a hexagonal arrangement in the earliest stages of their development, at the edge of a wave of morphogenetic activity, spreading out from the neural tube region and initiated afresh by each succeeding lateral row.

CHAPTER ELEVEN

MORPHOGENESIS OF A COMPLEX ORGAN: THE VERTEBRATE LIMB

THE DEVELOPING LIMB IS AN ATTRACTIVE SYSTEM FOR MORPHOGENETIC studies. It is much less complex than the whole embryo, but it exemplifies all of the fundamental developmental problems. In its early stages it undergoes fairly simple but interesting transformations both of its overall shape and in the differentiating tissues—particularly the rudiments of the skeleton—within it. Because its normal shape and differentiated pattern is well defined, deviations from it produced by experiment are easily recognized and scored, making it excellent material for investigation into the control of form and pattern in development. Moreover, this control appears to be particularly sensitive to genetic or environmental disturbances, leading to the occurrence of congenital abnormalities which give research in this field an important medical significance.

The amphibian limb disc as a morphogenetic field

Modern work was initiated by R. G. Harrison's studies, made between 1900 and 1925, on the developing limb of urodele amphibians. In the axolotl *Ambystoma* the limb appears first as a small round hillock, the limb disc, and he showed that this disc behaved, just like many early embryos and using the terminology invented by Driesch in 1891 to describe them, as harmonious equipotential systems. If half of the disc was removed, the remaining half produced a normal limb; a vertical division into two halves gave two complete limbs; and fusing two discs together produced a single limb, at first very large, but subsequently regulating to the normal size. These discoveries indicated that this structure is a true morphogenetic field, i.e. a region within which development of the parts proceeds in relation to their position in the whole, and therefore with a capacity for regulation if the size of the field is reduced or increased. Moreover, at an early stage when the limb disc is not distinguishable morphologically, the limb field is larger than the presumptive limb area, thus, if the whole disc is removed, a complete limb will still be formed from peripheral cells which would normally produce flank. As development goes on, the size of the limb field becomes steadily reduced until finally it coincides with the presumptive region from which the limb normally develops; the disc area is then determined as limb-forming, and the surrounding cells as flank.

The limb bud arises from a thickening of the lateral plate mesoderm (though Raynaud has shown that in reptiles cells from the somites contribute to the limb musculature, and this may be true of other vertebrates) together with its overlying ectoderm. It is the mesoderm which carries whatever factors determine its morphogenetic properties, since mesoderm implanted under flank ectoderm will lead to production of a limb at that site but a graft of the ectodermal component over flank mesoderm will not. In 1931 Rottman produced chimaeras by making interspecific grafts of the presumptive limb mesoderm in early gastrulae of the newts *Triturus cristatus* and *Triturus taeniatus*. These species have clearly distinct limb characteristics, and in the chimaeras the type of limb produced was always that of the mesoderm-donor.

Determination of polarity in the amphibian limb

Harrison also made important studies on the development of polarity in *Ambystoma*, in experiments in which he removed the limb disc and replaced it in the embryo, on the same or on the contralateral side, in the presumptive limb or the presumptive flank position, and with or without rotation to give an inverted or reversed direction. He showed that the polarities exhibited by the limb structures, most clearly in the form of the digits, were established in two stages. First the antero-posterior (AP) axis is established; before this occurs reversal of the limb disc will give a normal limb, since the polarity is regulated to conform with the surrounding tissues, but after this, when the axis is determined, the surrounding tissues can have no influence, and the limb will develop with reversal polarity. At this stage polarity in the dorso-ventral (DV) axis is still labile, but this is determined slightly later.

This aspect has recently been reinvestigated by Slack who showed in 1976 that the polarity of the AP pattern is caused not by some sort of molecular polarization of the cells within the disc as Harrison surmised, but by a response of these cells to a signal emitted from a region of flank cells just posterior to the limb rudiment. It can travel in both directions and involves no polarity at the cell level, but in normal circumstances it sets up a gradient of positional information across the limb disc, leading to the

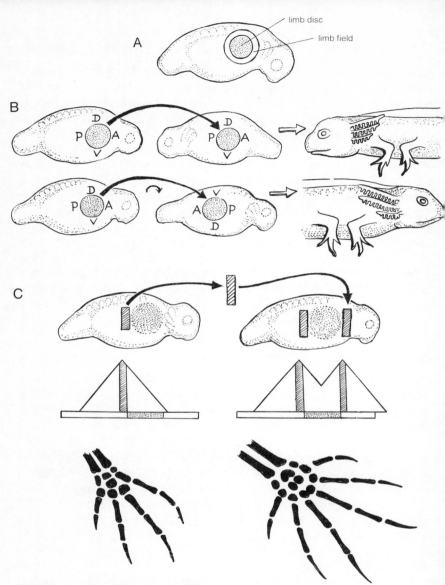

Figure 11.1 The limb field and polarity in the amphibian limb.
A. Distinction between the limb field (outer circle) and the limb disc (stippled) in *Ambystoma*.
B. Removal and regrafting of the limb disc indicates that antero-posterior polarity is irrevocably established before dorso-ventral polarity.
Based upon R. G. Harrison.
C. Duplicated limb parts with reversed sequence are produced when material from the posterior border of the limb disc (stippled) is transplanted to the anterior border in another embryo. The existence of a polar zone (hatched), setting up a gradient of positional information, is postulated to account for this.
Based upon J. M. W. Slack.

determination and differentiation of the various structures of the limb in their proper AP sequence. If a strip from this polarizing region is grafted anterior to the presumptive limb area, the disc receives signals from both directions; a double gradient is set up which leads to the formation of a limb consisting of two posterior halves arranged in mirror-image symmetry (compare "double embryos" in *Euscelis*, chapter 4). These findings explain the occurrence of reduplicated limbs which Harrison found very puzzling, particularly when it occurred in the case of a graft of a normally orientated limb disc to the flank, in harmonious relation (as he put it) to the rest of the body. His limb discs must have included some polarizing tissue, as well as presumptive limb material and, when grafted into the flank, they would continue to signal from behind, while the polarizing region of the host would in this experiment be signalling from in front, setting up a double gradient. This finding aligns this phenomenon with that produced by the presence of ZPA in the chick limb (see chapter 12).

Morphogenesis of the avian limb

Led by the work of Amprino, Saunders and Zwilling in the 1940s embryologists have turned increasingly to studies on the developing avian limb. It is testimony to the fascination of this system that so many investigators are now in the field, but to its complexity that so much basic information (especially at the cell level) remains to be discovered, and that none of the hypotheses regarding morphogenetic control mechanisms are established with certainty.

As the limb bud develops, the pattern of cartilage elements which prefigures the limb skeleton emerges: each element originates as a mesenchymal condensation within which chondrogenesis occurs, indicated clearly by synthesis of the acid mucopolysaccharide chondrotin sulphate which is the characteristic constituent, along with collagen, of the extracellular matrix secreted by cartilage cells. Development is essentially the same in fore- and hind-limb buds, with the most proximal elements appearing first and the most distal last. In the pentadactyl limb pattern (which is basic to all land vertebrates) there is a sequence of sets of elements along the proximo distal (PD) axis with an increasing number of elements in each set: a single one (humerus in the fore-, femur in the hind-limb) in the 1st set—forming the stylopod; 2 in the next (radius and ulna; tibia and fibula) in the next—forming part of the zeugopod; a complicated set of carpals and tarsals next—completing the zeugopod; 5 in the next set (metacarpals; metatarsals)—forming part of the autopod, 5 again in the

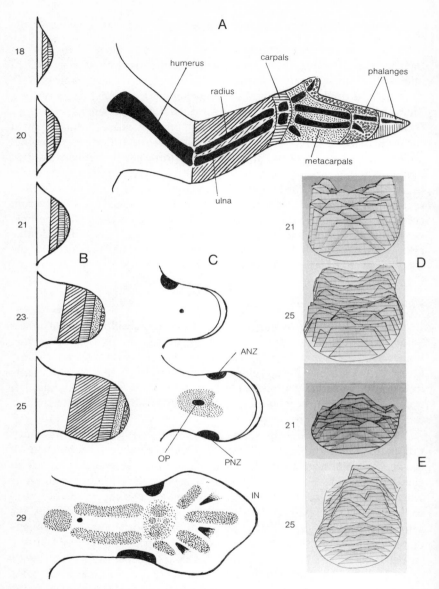

Figure 11.2 Normal development of the chick limb.
A. The wing of a 10-day embryo, showing the cartilaginous rudiments of the limb skeleton and the corresponding division of the limb into regions along the proximo-distal axis.
B. Growth of the wing bud (numbers refer to stages in the Hamburger-Hamilton table of normal development) with fate maps, showing the proximo-distalward sequence of development of the regions along that axis.

last set (phalanges)—completing the autopod. In the chick the number of digits is reduced to 4 in the wing and the leg.

The limb bud has an ectodermal covering which consists of two cell layers, an outer continuous layer of flattened periderm cells and an inner layer of columnar basal cells, rather loosely arranged and resting on a basal lamina—the ultrastructural layer present in most epithelia. The ectoderm is uniform, except at the distal edge of the limb bud, where it forms a thickened apical ectodermal ridge (AER), as Amprino has shown, by the ectodermal cells moving distally over the dorsal and ventral sides of the bud, and becoming compressed at the apex where they collide. The deformation of the cells at the collision line produce tall wedge-shaped basal cells, arranged fanwise in DV section. The AER is capable of reorganizing itself after disturbance, even, as Errick and Saunders showed in 1974, after the ectoderm has been turned inside out over the limb bud ectoderm. The mesoderm, which is the major component of the limb bud, consists in its early stages of a loosely packed uniform mesenchyme made up of cells of irregular shape, each one having many fine filopodial cytoplasmic extensions which make contact with neighbouring cells and, in the case of cells close to the ectoderm, with the basal lamina. During the course of development, the initial uniformity of the mesoderm cells is diminished as cell differentiation begins, to produce the distinctive histological structures associated with the skeleton, the musculature and the blood vessels.

One characteristic feature of development in the mesoderm is the occurrence of highly localized areas of cell death, easily visible in living embryos when these are stained with vital dyes. Some of these have an explicable role in development, notably the interdigital necrotic zones which appear between the digits. In the chick these lead to regression of the inter-digital tissue, but in web-footed species such as the duck cell death in this zone is much reduced. Much more puzzling are the anterior (ANZ) and posterior (PNZ) necrotic zones originally described by Saunders, Gasseling and Saunders in 1962. Each consists of a well-defined area situated immediately beneath the ectoderm behind the expanded distal tip of the limb at the anterior or posterior margin, produced by a wave of cell death which travels distally as the bud grows in length. It was orginally proposed

---

C. The appearance of mesenchymal condensations and cartilage rudiments of the skeleton (stippled) and of regions of cell death—the anterior and posterior necrotic zones (ANZ and PNZ), the opaque patch (OP) and the regions of interdigital necrosis (IN).
D. and E. Distribution of mitosing cells (D) and of cells per unit area (E) in the wing bud at Stages 21 and 25. By courtesy of M. B. H. Mohammed.

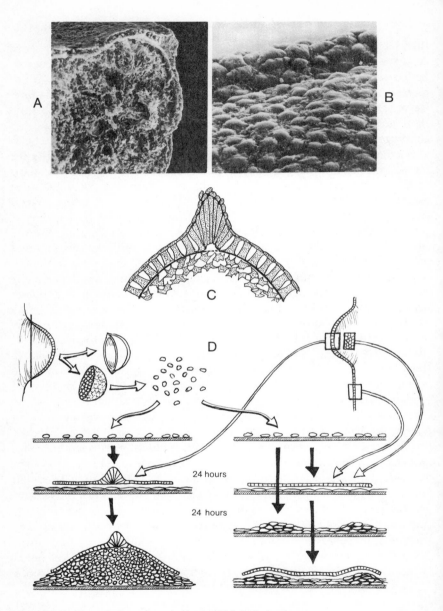

Figure 11.3 The apical ectodermal ridge in chick limb development.
A. Stereoscan electron-micrograph of a wing bud, sectioned along the proximo-distal axis in the dorso-ventral plane, showing the apical ectodermal ridge (AER), non-ridge ectoderm, and the mesoderm.

that these zones sculpted out the characteristic limb contours, but the cell death can be prevented without any effect upon normal morphogenesis. Saunders and Fallon showed in 1966 that this was to be regarded as a particular type of cell differentiation, with a pathway of determination like any other. Cell death was determined from the earliest appearance of the limb bud (Stage 17) though no cell death occurred until much later (Stage 24); presumptive necrotic zone cells grafted over somites died on schedule at Stage 24. But determination was reversible under certain circumstances between Stages 17–23, since if the cells were grafted in the dorsal side of a wing bud they remained healthy. No reason can be given for the occurrence and precise programming of these necrotic zones, and the cell death may be only an incidential manifestation of a control process affecting some more-significant event.

Cellular activities involved in limb morphogenesis

The cellular activities which together produce the change of form and pattern of the type found in limb development include proliferation by mitosis, cell death, cell movements and orientations, and the changes which occur in the course of cell differentiation to give the various tissue types— changes in size and shape, and also changes in packing density through secretion of extracellular material or other causes.

A mitotic gradient, with a high point distally, has been noted by several workers, but this is largely accounted for by cells ceasing to divide as they differentiate as cartilage proximally, and it is not in itself sufficient to account for limb outgrowth. In early stages all of the mesoderm cells are dividing, and at approximately the same rate in all parts of the limb bud. Cell death, as we have seen, plays only a small part in shaping the developing limb. The role of cell movement is less certain. Its occurrence cannot be observed directly for technical reasons, but where there is a predominately undirectional outgrowth, with no local distributions of

---

B. The AER and neighbouring non-ridge ectoderm at higher magnification. By courtesy of R. Bellairs.
C. Diagram of the periderm and basal layers of the ectoderm, and their modification at the AER. The basal layer rests upon a well-defined basal lamina, to which underlying mesoderm cells are attached by filopodial connections.
D. Culture of disaggregated limb mesoderm cells as monolayers leads to the differentiation of low cartilage nodules when no ectoderm is added or when non-AER ectoderm is added. Addition of AER ectoderm leads to heaps of cells up to 20 cells deep beneath the ridge and no cartilage nodules appear.
Based upon M. Globus and S. Vethamany-Globus.

mitosis to account for it and with no orientation of mitotic division, it must be presumed to occur. The difficulty is to establish its extent, and whether it involves active movement of cells over each other, or simply rotating around each other, or whether such movement as occurs is merely passive—imposed by forces such as the proliferating cell mass being constrained to expand in a particular direction by the enclosing ectodermal covering. Stark and Searls established in 1973 that at least there is no cell dispersal and intermingling of cells, since labelled mesoderm fragments elongate but do not otherwise lose their original conformation when grafted into a limb site. This does not eliminate the possibility of cell movement, but requires that it should be limited and/or orderly, and Ede and Law showed in 1969 that a good approximation to normal limb outgrowth could be generated in a computer simulation using a PD gradient together with some limited distalward cell movement. That the movement might be actively directed towards the AER is further suggested by observations by Globus and Vethamany-Globus in 1976 on limb mesenchyme cells in culture which aggregate beneath apical ectodermal ridges placed in the culture.

Ectodermal-mesodermal interactions

Amprino in 1965, and Hornbruch and Wolpert in 1970, have emphasized the biomechanical role of the ectorderm, especially the AER, in moulding the growth of the underlying mesoderm. Saunders in 1948 and subsequently Zwilling maintained that the ridge plays a much more active inductive role in limb outgrowth: when the ridge is removed, further outgrowth ceases; when an extra ridge is grafted to a limb bud, an additional limb grows out. Amprino explains the first result as due to the loss of the ectodermal covering which acts as a protection against destructive effects on the extraembryonic environment, and the second as resulting from providing an additional biomechanical mould. But the inductive effect appears to be limited to apical ridge ectoderm and, though Amprino has obtained pseudo-ridges in other regions of the limb bud, the structures which arise as a result undergo only a very limited development. Saunders and Zwilling also suggested that the mesoderm produced an apical ectodermal maintenance factor (AEMF) which was required for maintenance of the AER in a healthy state, and whose quantitative distribution within different regions of the bud determined the height of the overlying ectodermal ridge; insertion of an impermeable membrane between ridge and underlying mesoderm led to degeneration of the ridge.

That there is interaction between the AER and its underlying mesoderm is clearly established, but whether this involves synthesis and response to diffusible factors on both sides is much less certain.

Pattern determination along the proximodistal axis

In 1948 Saunders made the first presumptive fate maps of the presumptive areas of the developing chick limb by inserting carbon particles into recorded positions in the early bud and noting what structures they appeared in later. He showed that the regions appeared in proximodistal order, i.e that in the early limb bud the presumptive humerus region occupied most of the bud, then the presumptive radius and ulna were added, and finally the digital region. More recent work by Searls, using labelled grafts as markers, has shown that all of the presumptive regions are present from the earliest stages, but the more distal regions are compressed into a very narrow band, only a few cells deep, at the distal tip. Saunders further showed that removing the AER stopped further limb outgrowth, as described above, but more precisely it stopped outgrowth of those presumptive regions still undetectable by his method, i.e. those in the narrow distal band; those which were established as detectable areas did continue to grow and produce a limb stump, the extent of the stump depending entirely on how much of the prospective limb had been laid out as clear presumptive regions. The role of the AER was therefore not merely to induce outgrowth, but to promote the establishment of presumptive areas in PD sequence through interaction with the underlying mesoderm. This suggested that the AER had a capacity to induce specific regions at different stages (e.g. humerus induction by early ridge, digit induction by late ridge) but decisive experiments by Rubin and Saunders in 1972 showed this not to be the case; if the mesoderm is complete, the normal sequence of parts will develop whether it is capped with ectoderm from a younger or an older embryo.

A model for control of the proximodistal pattern which fits these facts well was proposed by Summerbell, Lewis and Wolpert in 1973. In this the function of the ridge, apart from its biomechanical one, is solely to maintain cells in the mesoderm underlying it in a labile undetermined state. Within this progress zone the cells undergo a change of positional value with time, which might be controlled by the number of mitoses each has undergone. Since the width of the zone is constant and the number of cells is continually increasing, cells are left behind as the bud lengthens and, as soon as each cell leaves the zone, its positional value is frozen as positional

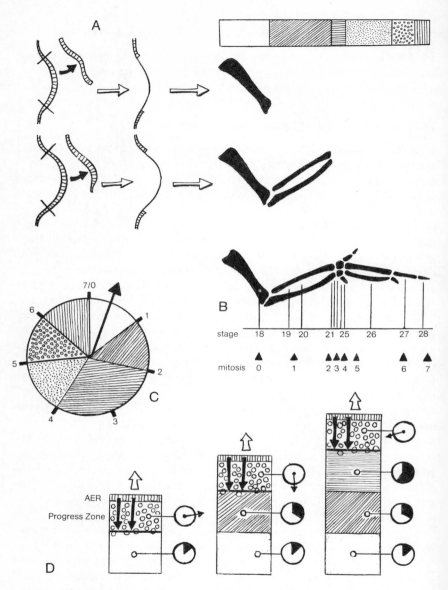

Figure 11.4 The progress zone model of the determination of limb parts in the proximo-distal axis.
A. Some experimental evidence: truncation of the limb following removal of the apical ectodermal ridge at different stages of bud development. The average level of truncation for each stage is indicated in B, together with the number of mitotic divisions each cell will have undergone at that stage.

information, in accordance with which the cell's developmental fate—as zeugopod, stylopod or whatever—is determined. According to this model, determination is completely autonomous, with no cellular interaction beyond (possibly) near neighbours, and therefore there should be no capacity for regulation of defects or excisions outside of the progress zone. Experiments by Kieny in 1964 suggested that such regulation does occur, but the point is difficult to establish beyond dispute. If it does, then the model proposed by Slocum in 1975, based on a positional informational gradient with the AER causing a very steep rise in the subridge region, would fit the facts better.

Pattern determination along the antero-posterior axis is described in chapter 12. The zone of polarizing activity (ZPA), which is the key to its control mechanism, was discovered by Saunders and Gasseling in 1968 in the course of studies on the posterior necrotic zone, with which it is spatially partly coincident.

Chondrogenesis in the developing limb bud

The study of differentiation and determination in cartilage development is an immense research field, and though most earlier studies were made on chick somite cells, either in the embryo or in cell culture, attention has turned increasingly to chondrogenesis in the limb bud, regarding which there are currently two theories. The first, maintained by Searls and others, holds that there are three phases of chondrogenesis: in the first, all mesoderm cells synthesize chondroitin sulphate; in the second, cells begin to be visibly differentiated—in the presumptive cartilage regions chondroitin sulphate synthesis is increased, but this is a reversibly determined state; in the third phase, when the cell's fate is irreversibly determined, the differentiation of the chondrogenic cells becomes visibly obvious through secretion of their characteristic cartilage matrix. Determination of cells as cartilage or muscle depends upon their position in the limb bud.

The second theory, advanced by Dienstman, Biehl, Holtzer and Holtzer in 1974, is based upon observations on clonal cell cultures of limb bud

---

C. Positional information might be obtained by the cell by reference to a mitotic clock, completion of each mitotic cycle corresponding to a particular limb region.
D. A factor produced by the AER determines the depth of the progess zone of mesoderm beneath the ridge. Cells in the zone are undetermined, but as soon as they pass out of it they are determined to undergo differentiation appropriate to one region or another, depending on how long they have spent in the zone, i.e. according to the positional information they receive at that time.

mesoderm cells, in which mixed clones of myogenic and chondrogenic cells were never found. This suggests that the early mesoderm cells are not all alike, but are already determined as cartilage or muscle precursor cells at least two cell generations before their cell lineages show any overt differentiation. On this view, which is at present a minority one, production of cartilage or muscle in particular regions of the limb, or in particular culture conditions, would depend upon selection of one set of cell lineages rather than the other to survive and replicate. This theory receives considerable support from the strong evidence obtained by Christ, Jacob and Jacob in 1977, based upon quail-chick embryo limb-chimaera experiments, that what Kieny has for long maintained—that the musculature is derived independently from the somites—is correct.

There is considerable evidence from studies of somitic cultured cells, e.g. those of Abbot and Holtzer in 1966, that physical interaction through cell crowding is necessary to induce and maintain cartilage differentiation. The formation of the mesenchymal condensations in the limb bud may represent an actual aggregation of cells as an initial step leading to chondrogenesis. Toole suggested in 1972 that the occurrence of hyaluronate before condensation, and its removal by hyaluronidase in the early stages of chondrogenesis, may be signficant in this connection; that the primary effect of hyaluronate was likely to be upon the cell surface, acting there as a regulator controlling cell interactions by inhibiting aggregation. If such an aggregation does occur, the cell movements would be very small, but their cumulative effect might account for the characteristic condensation pattern mentioned in chapter 1.

CHAPTER TWELVE

FORM AND PATTERN

From the genetic programme to patterns of determination

WE HAVE SEEN HOW DIVERSITY AND COMPLEXITY ARISE AND INCREASE in the course of development, and some of the events which bring this about, especially the processes of determination through which cells become committed to particular pathways of development leading to their overt differentiation as cells of one or another distinct histological type and to more subtle differences, e.g. of growth characteristics, within those types.

But of course the emergence from all these processes of an organism capable of survival and reproducing all the essential features of its species depends upon a precise allocation of these determined states in particular arrangements or patterns, and the transformation of these patterns into the form of the germ layers, rudiments, organs and overall morphology of the embryo through changes in the structural activity of the cells according to that pattern. Changes in form may have the effect of creating new patterns, which in turn lead to new cell activities and new forms, and so on, until determination is complete. Pattern and form are not merely the terminal result of embryogenesis—the zebra's stripes and the elephant's trunk—but exist all through development, simple at first, and then more and more complex. There is very little evidence that pattern ever emerges through unscrambling of cells determined in a random way (the case of cells which migrate over long distances—especially that of the melanocytes which produce pigmentation patterns—is a special case) and the phenomenon of cell sorting which follows mixing of different cell types seems more likely to represent an experimental aspect of that mechanism which restrains cells from dispersing in normal conditions, any breakdown of which, as in cancer, is likely to be disastrous for the organism; nevertheless, cells are themselves active individuals, not inert building blocks, and the way in which they make and break contact with neighbours, and shift and orientate in relation to them, may be crucial.

An approach through automata theory

How pattern and form are established is the most fundamental problem in developmental biology, upon which studies in genetics, the biochemistry of differentiation, cell interaction and morphogenesis converge. Embryogenesis consists of the implementation of a developmental programme of which each cell carries a complete copy encoded in its nuclear DNA; which parts of the programme are active in any cell at a particular time depends upon the local conditions in which the cell finds itself, and the history of its own activity and interactions, all of which determine its cytoplasmic state whose physical basis lies in its changing macromolecular composition and organization. How then do inherited patterns of cell determination arise?

In the case of mosaic development the answer is clear: the cytoplasmic states of cells in different regions of the embryo are different from the beginning, since cells come to contain distinct molecular components, distributed in a particular way in the egg. At the other extreme is the type of development in which all cells are initially in the same cytoplasmic state. This appears to be the case in mammals, where the cells of the morula are essentially identical, and diversification begins with their determination as trophoblast or inner cell mass (i.e. embryo proper) according to whether they are on the outside or in the interior of the spherical mass of cells. This type presents the problem in its most extreme form, the theoretical interest of which has led mathematicians such as Turing and von Neumann to provide the foundations of an abstract solution in terms of automata theory, the embryological implications of which were clearly set out by Arbib in 1972. This branch of mathematics is fundamental in the design of computers, and its application in developmental biology has led to the use of technical terms borrowed from that field: programmes, subroutines, states, memory, inputs and outputs, switching circuits and so on. An automaton is any device which receives a string of inputs, computes on this string and produces a string of outputs which will thereby change the state of the system and also modify the next string of inputs. In the embryo each cell functions as an automaton, whose macromolecular machinery is the genome and the genetic control mechanism in the cytoplasm; inputs consist of signals received from neighbouring cells or the immediate environment, but also of "memories" of previous outputs; outputs consist of changes in the cell's determined state and the synthetic and other activities characterizing its overt differentiation. Automata theory makes a valuable contribution in suggesting how a cell embedded in the embryo and receiving cues only from its immediate surroundings can make its proper contribution to the development of the embryo as a whole, contributing to the generation of staggeringly complex structures through essentially simple switching circuits such as those proposed by Kauffman in 1973 for controlling transdetermination in insects (see chapter 15) and the more general models for determination proposed by Wolpert and Lewis in 1975, where commitments to pathways of determination are registered ("remembered") as combinations of binary switches in biosynthetic feed-back circuits. Since it is impossible with existing biochemical techniques to analyze these intracellular molecular circuits directly, the cell remains in this respect a "black box", its contents unknown, and such models can only be based upon analyses of inputs and outputs, especially the cells' response through differentiation to changed genetic or embryonic situations; but they are none-the-less important in indicating what possibilities exist and what directions further research might take.

Developmental signals and cell interactions

Emphasis is currently focused chiefly upon the signalling mechanisms which provide the external cues for the embryonic cell's developmental activity. In many embryos, probably the vast majority, and we have seen examples in the sea urchin and in *Euscelis*, there is no detailed system of regional plasms in the egg or early embryo, yet neither is there complete cytoplasmic homogeneity; there is instead some system consisting of a few reference points to which developmental events throughout the embryo are clearly related. Even where such reference points do not exist initially, they must somehow be set up for development to proceed, and we have seen how this may occur in chapter 3. In either case, embryogenesis proceeds through interaction between cells which may act variously as the source, destroyer, transmitter or receiver of developmental signals, which will generally be molecular, leading to development of the various regions of the embryo in harmonious relation with each other.

The most direct sort of interaction is embryonic induction, already considered in chapter 9, which occurs between groups of cells, one or both of which are often in the form of an epithelium and one of which—referred to by Spemann as the *organizer*—provides a signal to which the other responds by entering a particular determination pathway. The earliest inductions are generalized, e.g. the so-called primary organizer in amphibians induces the whole neural tube and other axial structures, but these are followed by a hierarchical sequence of organizers with increasingly restricted regions of activity, e.g. the optic cup developed from the anterior part of the brain induces the lens, which in turn induces the cornea, while the otic vesicle developed from the posterior part induces the otic capsule, which induces the tympanic membrane. The subdivision is accompanied by a parallel contraction of the competence of cells to respond, as the pathways of determination open to them become increasingly restricted. This hierarchy therefore determines the location of many embryonic structures—once the position of the primary organizer is established, secondary organizers arise in relation to that, and tertiary organizers in relation to them—but it is inadequate as an explanation of the origin of precise and detailed patterns of development.

The existence of an inductive system implies a capacity for regulation in the sense that cells whose presumptive fate was to differentiate in one way will, if the organizer is displaced, give rise to another. But, as we have already seen, regulation goes further, so that embryonic fields are established such that after deletions within a field or fusion with a similar field the original pattern of determination is still produced. These fields are generally established within single sheets of cells and at their first appearance measure no more than about 1 mm (approximately 50–100 cells) across, though of course the pattern may not become visible until a great deal of expansion by cell division has occurred. The competent tissues responding to inductive stimuli are organized as fields, and the hierarchy of inductive interactions is therefore given precision by an accompanying subdivision of fields in the course of development; at first the whole embryo axis is a field, then as competence contracts subfields appear within it, e.g. the fields of the eyes, the ears and the limbs, within which new subfields emerge until all cells have entered their terminal determination pathways.

Fields, gradients and positional information

The characteristic of an embryonic field is that each part of it, probably each cell, develops in relation to its position in the whole field. We have already seen that in the case of the sea urchin and in the insect *Euscelis* regulative development exhibits the properties of a gradient or double gradient, and several embryologists, notably Child in 1928, suggested that gradients of some substance control development and account for size-invariant regulation within embryonic fields. But the gradients actually observed, e.g. of susceptibility to toxic substances and staining with vital dyes, were clearly irrelevant to the problem, and interest in the subject declined until it was reintroduced in a very striking form in 1969 by

Wolpert, who coined the term *positional information* for its main conceptual innovation. This was, that there exists in embryos a special system, which may be quite basic and universally distributed in animal embryos, whose function is to inform the cells of their position, upon which information the cells act in deciding which of the pathways of determination open to them they should enter. Development is not, according to this hypothesis, a step-by-step sequence, but a series of 2-stage events— first the specification of a cell's position within a field as input, and then the interpretation of that positional information by the internal circuitry of the cell with some change in the determined state of the cell, sometimes resulting in overt differentiation, as output. The positional information will be provided in most cases (an exception is provided by the progess zone model for determination of regions in the proximo-distal axis of the limb) by the cell's response to a morphogen which is produced in a high concentration at one end of the field and present in a low concentration at the other (often, it is believed through synthesis at the one—the source— and destruction at the other—the sink—with free diffusion through the intervening cells) with in between a continuous gradient of concentration whose slope will vary in steepness according to the distance. An analogy would be: a line of people (corresponding to cells), each of whom is provided with a set of flags (corresponding to potential pathways of determination), a book of rules (corresponding to the genes and genetic control mechanisms) and a measuring rod (corresponding to some receptor system within the cell or at the cell surface); the person at one end of the line holds at shoulder level a pole which the person at the other end rests on the ground. Each person then measures the height of the section of the pole in front of him, consults his rule book and then waves whichever flag the rules tell him is correct for that height. If, say, 4–6 ft indicated waving a blue flag, 2–4 ft a white, and 0–2 ft a red, the overall pattern would be that of the French national flag, and this pattern would be produced no matter how many people were added to the line or were marched away from it. A 2-dimensional pattern could be established by the provision of a second gradient at right angles to the first, thus providing in effect a grid of positional information, and a third dimension in the same way, though probably this would not often be necessary. A feature of the model is that the same grid may be used in different situations, e.g. in the wing and the leg, or in different species, but the cells in each case would respond differently.

The chief evidence for the existence of such a system has come from studies on regeneration in the freshwater polyp *Hydra* by Wolpert, Clark

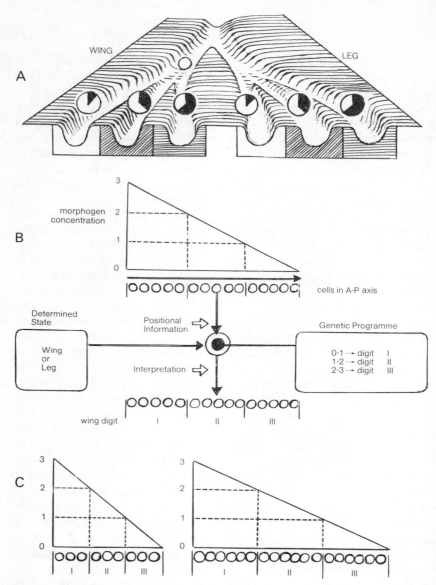

Figure 12.1 The concept of positional information.
A. The epigenetic landscape model applied to differentiation of the proximo-distal limb regions, with the selection of pathways of determination (limited to three in this diagram) decided by reference to positional information, provided according to the hypothesis depicted in Figure 11.4 by the cell's mitoitic clock. An earlier point of divergence decides whether the limb should be wing or leg, and the same positional information for determination of the regions will be used in each case; thus "upper limb" (stylopod) will be

and Hornbruch in 1972, and from the studies on insect cuticle patterns described in chapter 15, but its chief features are illustrated more simply in an analysis of the pattern of digit formation along the antero-posterior axis of the chick limb bud. In 1968 Saunders and Gasseling showed that a small region of mesoderm at the posterior edge of the limb bud, roughly coincident with the posterior necrotic zone and subsequently named the zone of polarizing activity (ZPA), if grafted either to the tip of another bud or to a position on the anterior edge corresponding to its normal position on the posterior edge, led (1) to the production of supernumerary limb parts and (2) to the imposition of a polarity in the arrangement of those parts—normal antero-posterior polarity in the case of the graft at the tip and reversed in the case of the anterior graft. The posterior elements of the additional wing structure were always formed adjacent to the graft, so that in the case of the anterior graft a wing with duplicated distal parts was produced with mirror-image symmetry, often with apparent fusion of digits at the mid-point. The most clearly developed digits in the chick wing are digits II, III and IV; sequences in the mirror-image duplications were therefore IV, III, II, II, III, IV or IV, III, II, III, IV. An analysis by Tickle, Summerbell and Wolpert in 1975 of a series of ZPA grafts made at all points along the A-P length of the bud showed these results to be consistent with the positional information model. The ZPA is, according to this hypothesis, the boundary or reference region for the antero-posterior axis of positional information. Since, according to the progress zone model described in chapter 11, only cells in the progress zone can respond to the influence of the ZPA, the character of the digits might be specified by their distance from the ZPA at the time when they are leaving the progress zone. A way in which the gradient could be set up which is consistent with the data would be for the ZPA to be the source of a labile morphogen which would act as a signal; in this case there appears to be no localized sink. A suitable gradient—of exponential instead of linear form—would be established if the morphogen were broken down outside of the ZPA region.

---

    interpreted as "humerus" in a bud determined as wing and as "femur" in a bud determined as leg.
B. Positional information provided by a gradient of morphogen concentration, exemplified in the determination of the order of digits in the antero-posterior axis of the limb (see figure 12.2). The cell *(centre)* is regarded as an automaton, with inputs of positional information and "memory" of previous decisions (leg or wing) and a genetic programme providing instructions which will decide the pathway of determination to be entered next.
C. Regulation occurs in this system because the positional information will lead to the same proportional division of the axis irrespective of its length.
Based upon L. Wolpert.

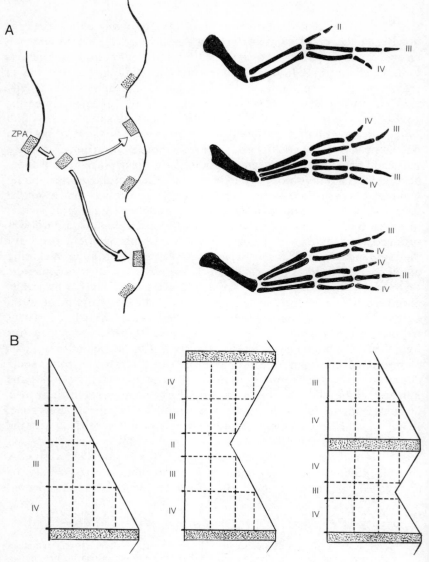

Figure 12.2 Determining the pattern of digits in the chick limb.
A. Results of transplanting material from the zone of polarizing activity (ZPA) of one wing bud to the anterior border and to the tip of another.
   Based upon J. W. Saunders Jr and M. T. Gasseling.
B. Interpretation of these results in terms of positional information.
   Based upon C. Tickle, D. Summerbell and L. Wolpert.

It should be mentioned, however, that in 1977 Saunders, who originally discovered the ZPA, summarized some doubts which experiments by himself and by Fallon and Crosby had thrown upon the necessity of its presence in normal development.

Polar co-ordinates in limb regeneration

The power of such an analysis to explain apparently mysterious results is well illustrated by work by French, Bryant and Bryant in 1976 on limb regeneration in insects and urodele amphibians, in which similar results are obtained. In the amphibian, for example, cutting of the distal half of the adult limb and rotating it through 180° before grafting it back on to the stump leads to the production of two supernumerary limb outgrowths at the graft junction. This is understandable if limb parts are specified through polar co-ordinates of positional information as illustrated in figure 12.3. The important co-ordinates here are those which form a circle around the cut surfaces of the limb, one of which will be displaced relative to the other. Two rules account for the supernumerary limbs: (1) growth occurs to intercalate missing positional values, between those defined on the two exposed circles, always producing the shorter set, e.g. if the value were 4 on the stump circle and 8 on the tip (the whole circle being 12), a line between would be established with values 5, 6, 7 but not 9, 10, 11, 12, 1, 2, 3; (2) outgrowth of a new limb occurs whenever a complete circular sequence of positional values is exposed by amputation or generated by intercalation. At two points on the graft boundary complete circular sequences are set up, since both sets are equal, e.g. between 3 and 9 they would be 4, 5, 6, 7, 8 and 10, 11, 12, 1, 2, and therefore both are generated; consequently the two supernumerary limbs grow out at these points, on either side of the original limb.

Alternative theories of pattern determination

There are a number of difficulties in the way of accepting positional information theory, at any rate as a universal mechanism: the separation of a special process of information signalling from a second process of interpretation and implementation; the necessity for a special and quite unknown mechanism for setting up special groups of cells as boundaries; the burden of complexity which is placed on the interpretive mechanism as the price for simplicity in the signalling system—these present problems for which other models, where more emphasis is placed on short-range

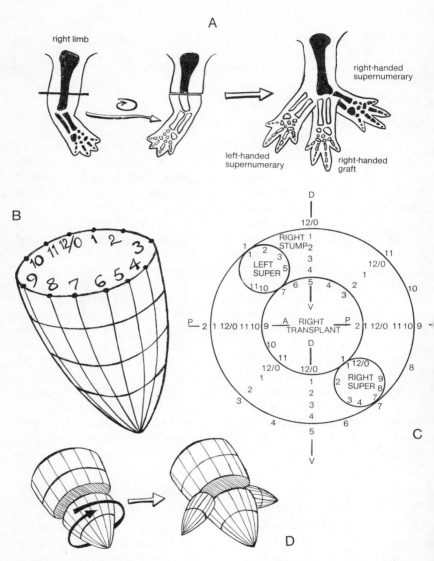

Figure 12.3 Positional information based on polar co-ordinates in regenerating limbs of adult amphibians.

A. The distal part of the limb is cut off and regrafted after rotation. Two supernumerary limbs grow out from the graft junction, one of them left-handed with reversed orientation relative to the other side and to the graft. Black indicates dorsal side.
B. The system of polar co-ordinates which provides positional information by reference to a grid consisting of radial co-ordinates which extend from a point at the tip of the limb and

interactions between cells, each of which has no information about its position relative to some distant and special group of cells but only about what its neighbours are doing and its own internal state, suggest alternative solutions. One of this type—the cell contact model—was developed by McMahon in 1973 and further in 1976, originally for regulation in the slime mould grex (see chapter 1) but which accounts for regulation in *Hydra* in a way which contrasts interestingly with the positional information explanation.

Periodic patterns in development

A special problem arises in the case of periodically repeated structures such as somites, for which a model based on an application of the extremely interesting mathematical "catastrophe" theory of how discontinuous and divergent phenomena arise in continuous systems, was developed by Cooke and Zeeman in 1976. In such systems structures are formed at regular intervals along an axis (e.g. the cartilage rudiments of the digits along the antero-posterior axis of the limb bud) to produce which the positional information model requires a multiplicity of thresholds, with a correspondingly complex biochemical switch mechanism sensitive to many triggering levels, whereas the same initial pattern may be produced using a wave-form gradient, with a single threshold intersecting its peaks. Such a system—sometimes called a *prepattern* because what is ultimately revealed as a differentiated pattern depends upon the variable level of the threshold against a stable underlying gradient, as the tide reveals different patterns of rocks on the seashore according to its depth—was invoked in an analysis of the arrangement of bristles in *Drosophila* by Maynard, Smith and Sondhi in 1961. They utilized, to generate the underlying wave form, a model proposed by Turing in 1952 in which two interacting morphogens produced, from completely uniform initial conditions and random perturbations, a stable chemical wave pattern. Bard and Lauder in 1974, using computer simulation techniques that were unavailable to Turing, demonstrated that this system might indeed account for the distribution of

---

circular co-ordinates which traverse these radials at intervals along the length of the limb. Intervals between the circular co-ordinates are actually irregular, as in C.
C. The experiment illustrated diagrammatically to show the displacement of these co-ordinates relative to each other, and the establishment of new ones to complete the sequences. For clarity, the diameter of the graft is made smaller than that of the stump.
D. The results in terms of this model. For further explanation see text.
Based upon V. French, P. J. Bryant and S. V. Bryant.

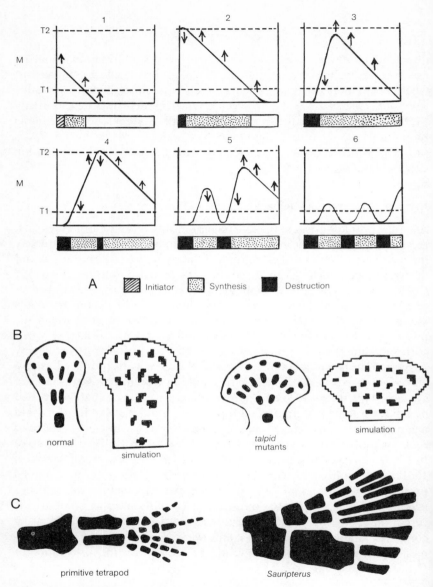

Figure 12.4 Periodic patterns and the development of cartilage patterns in the vertebrate limb.

A. A model which sets up a stable wave-form gradient which might underlie periodically occurring structures such as the digital cartilage condensations along the antero-posterior axis of the limb. A morphogen (M) is synthesized, firstly by cells in a special initiator region, and diffuses along the axis where other cells respond to it in the following way: at concentrations of M greater than threshold T1 cells synthesize M; at thresholds greater epidermal hairs and bristles, but was not sufficiently precise to give the pattern of digits and similar structures.

In 1975 Wilby and Ede proposed a way in which a stable wave-form gradient could be set up to produce the initial pattern of skeletal cartilage differentiation in the limb bud, using only localized cell-to-cell interactions, initiated in the ZPA, in which cells modified their metabolism irreversibly at critical threshold levels of a diffusible morphogen which could be synthesized or destroyed by every cell in the system. The pattern of cartilages which emerges depends upon the shape of the developing limb bud, the position of the initiator region, and the rate of synthesis of the morphogens. Thus a change in the shape of the developing limb bud would be expected to alter the number of digits, with a broadening of the limb bud producing polydactyly; this is what does occur in the *talpid*^{3} mutant where, as we shall see in chapter 13, genetic effects upon cell adhesion and movement lead to the production of a fan-shaped limb bud, with about twice the normal number of digits. The positional information model would produce a size-invariant pattern by altering the relative size or spacing of the digits, but not by adding any, and in this respect the waveform gradient appears to fit the facts better. On the other hand, the growth characteristics of the cartilage rudiments are clearly defined, such that in organ culture each isolated rudiment will go on to produce its normal morphological form, with all the bumps and knobs that define it as humerus, thumb or whatever, and this appears to require some specification which is most easily accounted for by a positional information gradient, which might, however, be superimposed on the periodic pattern of cartilage differentiation. A condensation process leading to the formation of cartilage nodules occurs in mesenchyme cells cultured as aggregates or as sheets, producing simple distribution patterns, and it may be that the tendency of cartilages to arise in periodic patterns, given further ordering through the developing shape of the limb, has produced the most

---

than T2 cells actively destroy M; the transformations from inactive to synthesizing and from synthesizing to destroying are irreversible. In this case the destructive phase is linked to the formation of the digital condensations in the mesenchyme.

B. The model predicts that, other parameters being equal, the cartilage pattern will be a function of limb shape. Computer simulations show that parameters which give the normal number of digits in a normal limb-bud will produce an excessive number of digits in a fan-shaped limb-bud, and this is found to be the case in *talpid* mutants of the fowl (see figure 13.3).

C. Comparison of the pentadactyl limb of a primitive land vertebrate, with the fin of a Devonian crossopterygian fish *Sauripterus*.
Based upon O. K. Wilby and D. A. Ede.

ancient evolutionary limb patterns, which have led to more complex patterns through the superimposition of positional information systems as a sort of fine-tuning. In the *talpid*³ mutant, and in limbs produced from reaggregated mesenchyme cells, the fine-tuning appears to have broken down, resulting in a limb pattern which is more like that of the crossopterygian fish fins from which the tetrapod limb is thought to have evolved. How far this fine-tuning is carried, and how far development depends upon each cell or small group of cells being assigned a particular state in the course of determination, on which its future developmental activities depend, is an important problem which arises in its most striking form in the development of the nervous system, discussed in chapter 17.

CHAPTER THIRTEEN

GENES AND DEVELOPMENT

TO UNDERSTAND THE RELATION BETWEEN AN ORGANISM'S GENOTYPE, ITS genetic constitution, and its phenotype, the result of the activities of its genes, is one of the chief aims of developmental biology. The phenotype may be considered at any level, from the cellular and subcellular up to studies on organs or whole organisms; in any case it is not a static thing but something which changes in the course of development as the cells move down the pathways of the epigenetic landscape—or sometimes, as we have seen, back up them and into different ones.

Development and evolution

Ultimately the evolution of organisms must be considered in these terms, since natural selection acts upon the phenotype but produces its results through changes in the genotype, and any evolutionary change must be such as the developmental mechanisms will generate through modifications of existing developmental processes. Some modifications will be much more likely to occur than others, and some will set up systems where other modifications are an almost inevitable consequence, just as one human invention suggests another, leading to an appearance of directed evolution along particular pathways. The changes of body form produced as a result of differential growth and analyzed by D'Arcy Thompson in 1917 in his classic study *On Growth and Form* by the method of Cartesian transformations, are of this type. The form of the structure is defined by superimposing a grid and deforming the co-ordinates of this grid (how this is done in systems consisting of many individual structures is quite unknown, and is one of the major problems of growth control); it produces a series of topologically identical but otherwise different forms, the extremes of which, e.g. the skulls of the primitive reptile *Dimorphodon* and the extinct flying reptile *Pteranodon* shown in fig. 13.1 appears to be totally distinct.

156 GENES AND DEVELOPMENT

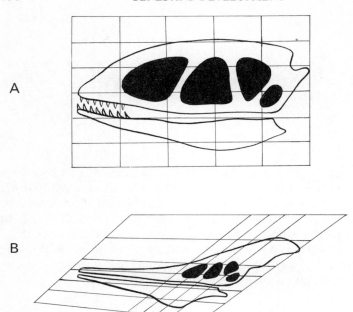

Figure 13.1 Cartesian transformation showing how the skull of *Pteranodon* (B) may related to that of *Dimorphodon* (A).
Based upon D'Arcy Thompson.

One especially interesting aspect of developmental biology in this evolutionary context is the problem of accounting for adaptive features of the type which are produced generally as a reaction to an environmental stimulus, but which in particular cases appear as inherited characteristics. For example, as we saw in chapter 10, calluses—extreme thickenings of the stratum corneum—appear whenever the skin is subjected to rubbing; but on the sole of the foot and—more unexpectedly—in patches on the breast and rump of ostriches, in areas which press against the ground when the birds are squatting, these calluses appear during embryonic development. Lamarckian explanations are nowadays discounted, yet random mutations for development of calluses in exactly the right place are almost as hard to accept. In 1957 Waddington proposed an explanation in terms of genetic assimilation, based upon his epigenetic landscape model and the canalization of development which it implied, i.e. the adjustment of the genotype to produce one definite result, regardless of minor genetic and environmental variations, His suggestion was that in the ostrich the response to the environmental stimulus in the ancestral ostrich was facilitated and made more uniform by selection, establishing a branch pathway which still required an environmental stimulus to shunt development into it; but the pathway being established, a wide variety of minor environmental or genetic stimuli would do it, so that it is very probable that a suitable random mutation would arise and be selected, so that the whole process would come under genetic control. Waddington showed that genetic assimilation of a comparable type could be demonstrated in the laboratory, using a stock of the fruit fly *Drosophila* in which a small proportion of the population responded to a heat shock by forming a wing lacking the usual crossvein. Selection for response to the heat shock produced a stock in which some flies showed the crossveinless condition without being exposed to the stimulus, and further selection from these individuals led to a stock in which crossveinlessness occurred spontaneously in most of the flies.

Genes in individual development. Pleiotropy

In order to discover how genes act, it is necessary to substitute mutant genes for their normal wildtype alleles and observe what effect this substitution has upon development; if embryos consisted of a mosaic of autonomous parts and processes, and if each gene affected only one part or process, it would then be easy to define the activity of the gene by simple subtraction. But we have seen that this is not the case; the embryo is a highly complex interacting system, and a genetic alteration in one part of the system is almost certain to have repercussions in other parts. A spurious suggestion of simplicity often arises from the names which mutants are given, referring to the character or structure which is most obviously identifiable anatomically in the fully developed embryo or adult. For example, a number of genes have been given the name "wingless" in the fowl because, indisputably, the birds have stumps or no wings at all, but the gene *wg* leads as well to absence of the lungs and air sacs, and also the metanephric kidney, suggesting that some widespread cell function is affected which is manifested only in regions of some specific sensitivity; it is not a gene concerned specifically with controlling the development of the wing. No simple "one gene-one character" relationship exists; generally a mutant gene will produce a multiplicity of different effects and this phenomenon is known as *pleiotropy*.

Therefore, the first stage in analysis of gene action is to establish what

Grüneberg, who surveyed a long series of studies on mouse mutants in 1963, has called a *pedigree of causes*, tracing each terminal manifestation back through its embryological development to its origin in some primary effect, from which all the abnormalities will also have arisen and diverged. Frequently this includes the analysis of inductive interactions which, having failed to occur, or occurring in abnormal locations through displacement of the interacting tissues, lead to a cascade of further developmental effects. One of the mutants in Grüneberg's survey illustrates how difficult such an analysis may be: mice homozygous for the gene *congenital hydrocephalus (ch/ch)* show widespread skeletal abnormalities and also bulging cerebral hemispheres. The primary effect of the gene appeared to be on the formation of cartilage condensations in the mesenchyme, producing the skeletal defects, including deformities of the skull which led mechanically to abnormalities of the brain which in turn interfered with the normal distribution of cerebrospinal fluid. But in 1970 Green showed that there were a number of other effects, especially urinogenital defects arising from an enormous excess of mesonephric (embryonic kidney) tubules which filled up the whole region between the normal mesonephros and the kidney, and there was also a striking deficiency of cells in the neighbouring coeliac nerve ganglion. These two are likely to be related, since neural-crest cells, from which the ganglion cells are derived, inhibit the development of mesonephric cells in normal mice. Green suggests, as a tentative unitary hypothesis which combines all of these aspects, that the *ch* gene may have an effect upon the mesenchyme which inhibits the movement of neural-crest cells through it and also interferes with cell interactions involved in the mesenchymal condensation process.

Hypotheses of this sort have only a limited value until they are supported by experimental evidence, of the sort which Mayer and Green in 1968, and Mayer in 1973, produced in analyses of the mutants *steel (st)* and dominant white spotting *(W)* in the mouse. The pleiotropic effects of the *steel* gene are anaemia, sterility and white coat colour, caused by the absence from the skin of melanocytes derived from the neural crest; the retinal pigment cells are pigmented so that these white mice do not have pink eyes. In all these cases the effect is not directly upon the cells which fail to develop, but upon the tissues in which their terminal differentiation normally takes place: primordial germ cells are present but very few arrive at or survive in the gonads; the precursor cells of the erythropoietic system are present and can be used to repopulate the bone marrow of irradiated normal hosts; the neural crest produces melanoblasts which differentiate into melanocytes when combined with normal skin, but normal melanoblasts in *st/st* skin produce no melanin, either in the epidermis or dermis, even when one or the other component is from a normal mouse. In the *dominant white spotting* mutant, which is also sterile and white, the reverse is the case; the effect is directly upon the cell, and its skin supports normal differentiation of melanoblasts from *wildtype* or from *steel* mice. Whether the inhibiting effect of the *steel* tissues is upon the migration of the cells through them, or upon their survival or their differentiation when they arrive at their destinations, is at present unknown.

The use of mutants as experimental tools

Apart from the information they provide about interacting systems in development and their genetic control, mutants may be very useful experimental tools when their effects on embryogenesis are clear. This is pre-eminently the case in the *nude* mouse mutant. Homozygotes *(nu/nu)* are hairless, stunted in growth and infertile, and in 1968 Pantelouris showed that all of these effects result from the failure of the thymus gland to develop from the epithelium of the pharyngeal pouches. In normal mice the thymus becomes populated with lymphoid progenitor cells derived from the multipotent haemopoietic cells which arise in the yolk sac; these undergo determination in the thymus as T lymphoid cells, which then migrate to other sites—the spleen and lymph nodes, as do other lymphoid cells which originate directly in the bone marrow and are then determined as B cells. Both types develop as lymphocytes, circulating through the body and acting in a synergistic way to produce an immunological response against foreign antigens. The *nude* mouse, since it lacks T cells, is equivalent to a perfectly thymodectomized animal, incapable, for example, of rejecting grafts of foreign tissues. The development of the immune system is too complex to describe here but, with its capacity for producing lymphocytes responding to specific antigens through cell-surface recognition mechanisms, its particular sort of determination which leads to a different response to the second from that to the first exposure to an antigen— a type of cellular memory—and its cells which may function as agents of immunological surveillance against potential tumour cells, it constitutes one of the most exciting and important areas in developmental biology (see, e.g. Goldschneider and Barton's review in 1976 of work on the development and differentiation of lymphocytes), and the *nude* mouse has provided an important research tool in this field.

The T-locus in the mouse

Wherever in development there is interaction between one cell and another, or a response by a cell to some factor in the intercellular substance, it will be mediated through some activity or property of the cell surface, e.g. through communication via cell junctions, response at receptor sites, or adhesions at points of cell contact. As we have seen, many genes act by modifying tissue interactions and in some cases analysis at the cellular level, and at the cell surface level, has begun. The long series of investigations of the *T*-locus by Bennett and colleagues, reviewed by Bennett in 1975, illustrates the direction such research is taking.

The *T*-locus is a genetically complex region of the 17th chromosome in the mouse, first recognized with the discovery of the dominant mutant *Brachyury (T)*, which produces a short tail in heterozygotes and embryonic lethality in homozygotes. Subsequently, a number of recessive alleles *(t)* have been discovered, which constitute a string of genes at the locus, each of which can be identified by giving a tailless phenotype when combined with the *T* gene. Five of these recessive alleles are also lethal as homozygotes, producing a specific type of developmental abnormality. In spite of this they are extremely wide-spread in wild population of mice, and for a very interesting reason: these genes are transmitted to their progeny in the normal Mendelian ratios by female heterozygotes, but males transmit lethal *t*-alleles in extraordinarily high ratios (75–90%). It is probably significant in this connection, since fertilization is so much a matter of interacting cell membranes, that specific cell-surface antigens determined by genes at the *T*-locus are found in sperm though not in other adult cells.

In 1964 Bennett showed that the lethal alleles affect a sequence of early developmental events involved in axial organization, at each of which there is some developmental switch or significant advance in the pathways of ectodermal/mesodermal differentiation. For example, in t^{12} development stops at the morula stage when the trophoblast should separate from the inner cell mass; in t^0 when the extraembryonic ectoderm is separated from the embryonic ectomesoderm; and in t^9 when a primitive streak is formed and the mesoderm cells migrate in and separate from the ectoderm. In t^9 embryos the primitive streak is extremely enlarged and there are very few mesoderm cells; the enlargement is caused by immobilization of the cells in the primitive streak and the failure of most mesoderm cells to invaginate in the normal way. The mutant mesoderm cells form tight clusters and resemble the primitive streak cells observed in earlier stages of normal embryos. It appears that they have not migrated into their appropriate

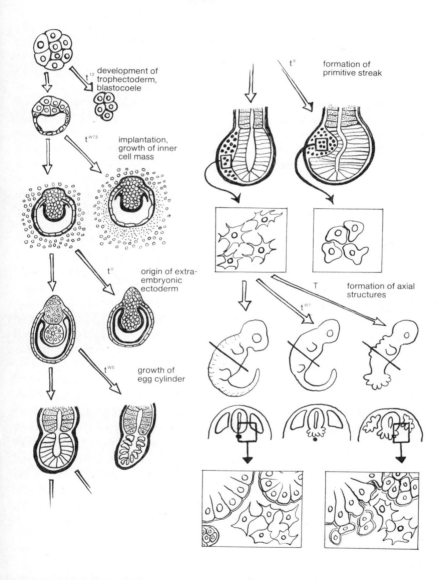

Figure 13.2 T-locus mutants in the mouse, showing the effects of different *t*-alleles on a series of stages in embryogenesis. (For details of normal mouse development see figure 16.2).

location and have consequently been prevented from completing their normal differentiation. The reason for the failure of normal movement is clear at the cell level, where light and electron-microscope studies show that whereas normal mesoderm cells are in contact with each other by fine filopodial processes, the mutant cells have broad blunt lobopodia, which make contact over large areas of their surface. This suggests that the primary effect of the gene is to produce some abnormality at the cell surface; electron-microscope analysis does indeed show that the system of microfilaments immediately below the surface is absent in t^9/t^9 cells, but this might be a manifestation of a surface defect. The result is that the whole gastrulation process is distorted, and a cascade of defects in all the axial structures follows—in the notochord, the neural tube and the somites. Studies on the other T-locus mutants, though not made in such detail, show that in all cases cells are unable to reach their normal locations or to survive if they reach them; but the different alleles cause this abnormality to appear at different stages in development. Bennett's hypothesis is that the role of the wildtype genes at the T-locus is to control the sequential expression, on specific cells, of surface components necessary to successive stages of differentiation, through interference with both morphogenetic movements and cellular interactions.

Serological evidence has been obtained which supports the suggestion that these genes are acting at the cell surface, using—since it is impossible to obtain embryonic cells in sufficient quantities—of the teratocarcinoma cell line F9. Teratomas are not only of great interest in themselves, but they provide powerful instruments of developmental analysis, as described by Stevens in 1967. They are embryonal tumours which arise from abnormal germ cells which continue to proliferate and differentiate in a number of ways: they may consist of multipotent stem cells which generate cells which differentiate along a variety of pathways; or they may give rise to lines of cells in which differentiation is restricted to one particular pathway; or they may produce lines which remain as undifferentiated cells. F9 is of this last sort, and its cells probably correspond to cleavage and morula cells in the embryo; Artz, Bennett and Jacob in 1974 made a good case for supposing that a cell surface antigen obtained from F9 is specified by the wildtype allele of t^{12}, which, as we have seen, is the mutant which does produce its effects at the morula stage.

The talpid[3] mutant of the fowl
For technical reasons, studies of cells in embryos can generally consist only of light and electron-microscope observations on fixed specimens. In-

vestigations of cells in culture are not subject to such limitations; the behaviour of cells as individuals, responding to the comparatively simple social system set up in the Petri dish, can be observed, analyzed by time-lapse film, and the activity of normal and mutant cells compared. Differences in the activity of normal and mutant cells in culture can then be related, by drawing inferences about what consequences these differences will have in the more complex embryonic situation, to the cells' normal roles in morphogenesis. This sort of analysis has been carried out by Ede and Flint in 1975 on *talpid*^{3}, a recessive lethal gene which in homozygotes *(ta^3/ta^3)* causes, among other pleiotropic effects (including the unique one of producing small ectopic crystalline lenses embedded in the head mesenchyme), development of broad fan-shaped limb buds instead of normal elongate paddle-shaped ones.

Mutant limb bud mesenchyme cells in monolayer culture move more slowly than normal cells, they adhere to each other and to the substrate more closely, and in confluent cultures an interconnecting network of vaguely-defined chondrogenic regions is formed rather than a system of well-defined circular condensations. Normal limb mesenchyme cells in culture are elongated, with a ruffled membrane at the leading edge; *talpid*^{3} cells are much more flattened, with short spiky cytoplasmic extensions all around the cells' periphery, which attach to the plastic and to other cells and tend to immobilize the mutant cell. In 1974 Ede, Bellairs and Bancroft showed in a transmission and scanning electron-microscope study that essentially the same cell morphologies and therefore, presumably, the same basic differences in the cell activity were shown by the cells in the limb bud. The probability that a slight distalward movement of mesenchyme cells is an important factor in limb bud growth was mentioned in chapter 11. It is impossible to observe such a movement directly, but in 1969 Ede and Law devised a computer simulation model in which some of the cellular activities thought to be important in limb bud growth were programmed in an elementary way. Proliferation with a proximo-distal gradient such as is known to exist, together with a slight distalward cell movement, produced a reasonable approximation to the normal paddle-shaped limb, but reducing the movement produced a broad fan-shaped limb, similar to that found in the *talpid*^{3} mutant. Thus the observed behaviour of the cells in culture, their morphological appearance in the limb bud, and the results of the computer simulation combine to suggest that the primary effect of the *talpid*^{3} gene is upon the capacity of the cell surface (microfilaments are normal) to produce cytoplasmic extensions associated with cell movement and cell contact, and that this reduced capacity for movement plays a large part in

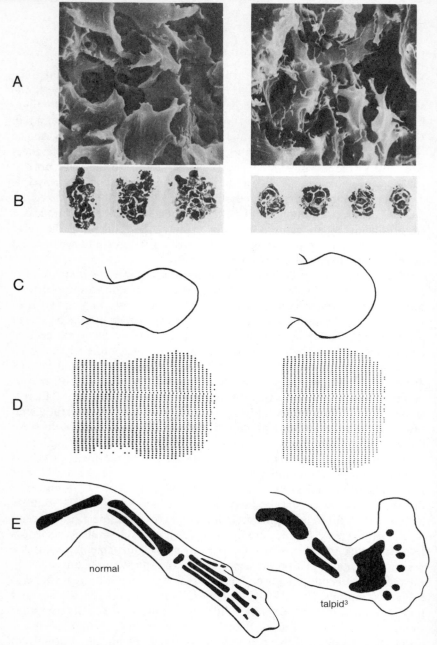

Figure 13.3 Cell properties and limb development in normal *(left)* and *talpid*[3] mutant fowl embryos.
A. Scanning electron-micrographs of early limb bud mesenchyme.

producing the characteristic distortion of the normal limb-bud shape. The pattern of skeletal rudiments within it is also highly abnormal, and this aspect of the *talpid*³ mutant has been discussed in chapter 12.

Gene control systems

Very little is known about the intracellular molecular mechanisms concerned with regulation of gene activity in higher organisms, but a system comparable to that established by Jacob and Monod in 1961 for bacteria, though necessarily of much greater complexity, was proposed by Britten and Davidson in 1969. There are no less than five classes of element in their model: sensor genes, integrator genes, activator RNA, receptor genes and producer (all those not engaged in regulating the genome; analogous to the structural genes of Jacob and Monod) genes, but its essential feature is that producer genes—those directly concerned with the synthesis of the enzymes and structural proteins involved in embryogenesis—are arranged in batteries, each of which is under the control of a single integrator gene. Evolution of complexity in evolution has been accompanied by an increase in the size of the genome (thirtyfold between sponges and mammals) which can hardly be accounted for in terms of additional producer genes required; complexity of structure and therefore of development would have occurred rather through a vastly increased complexity in a combinatorial network of regulatory genes (the integrator and receptor genes). In such a system a mutation of a regulatory gene would affect a whole battery of producer genes, leading to a multiplicity of embryonic disturbances and therefore to a type of pleiotropy which would not depend upon a cascade of developmental consequences following a single primary effect. It is very difficult to distinguish the two in practice, but the *Notch* mutant of *Drosophila* described by Poulson in 1945 produces such a wide range of disturbances in early development, apparently unconnected through embryonic interactions, that the normal gene at this locus must always certainly be of the regulatory type.

B. Aggregates produced in rotation culture from disaggregated limb mesenchyme cells.
C. Outlines of limb buds of 6-day embryos.
D. Computer simulations of normal and *talpid*³ limb bud development, in which only the amount of distalward cell movement is different.
E. The cartilage skeleton of legs from 12-day embryos.
Based upon D. A. Ede, R. Bellairs, M. Bancroft, O. P. Flint and J. T. Law.

CHAPTER FOURTEEN

HORMONAL CONTROL
OF DEVELOPMENTAL PROCESSES

Mammary gland development in the mouse

HORMONES ARE CHEMICAL SUBSTANCES WHICH CIRCULATE FREELY throughout the body, affecting the activity—in embryos, the differentiation—of a variety of target tissues which are often widely dispersed and whose responses to the same hormone may be quite different. They can therefore play no primary part in establishing patterns of morphogenesis, but act as switches—on or off—of gene expression in cells whose developmental options have already been narrowly restricted by prior determinative events. Thus, five pairs of mammary rudiments are produced in a specific pattern in all mouse embryos of whatever sex, but in males they regress in response to secretion of the androgenic steroid testosterone by the developing testes. Using organ culture techniques in 1971, Kratochwil showed that the genetic sex of the mammary gland has no influence on its developmental capacities—glands of male embryos are able to develop in the absence of testosterone, and those of females will regress in the presence of it. The rudiments appear at the 12th day of gestation, consisting (see chapter 10) of mesenchymal and epithelial components, the epithelium forming buds around which the mesenchyme condenses. In the next two days this forms a branching glandular system in female embryos, but in males the stalk of the bud becomes pinched off by a dense accumulation of mesenchymal cells, and the bud itself becomes smaller and necrotic, then is cut off from the stalk, and eventually disappears. In 1976, Kratochwil and Schwartz used a mutant—*testicular feminization (Tfm)*—to analyze the hormone's site of action. The *Tfm* gene is on the X-chromosome, and all male mice carrying it are insensitive to androgenic hormones—apparently through impaired function of the androgen receptors at the cell surface—so all their secondary sexual characteristics, including the mammary glands, are female in type. In reciprocal recombinations of normal and mutant components in organ culture, regression occurred as a response to testosterone when *Tfm* epithelium was combined

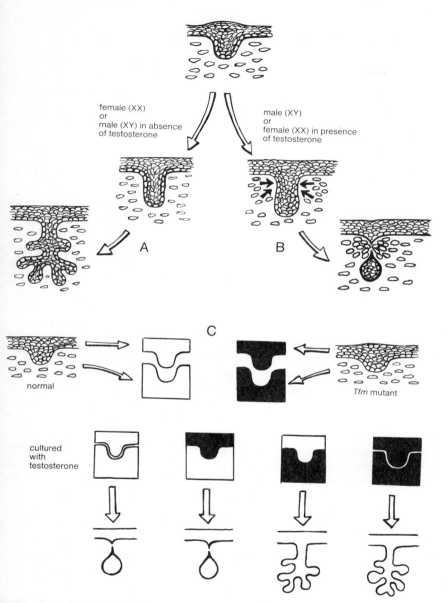

Figure 14.1 Hormonal control of mammary gland development in the mouse.
A, B. Development of the mammary gland rudiment in (A) female mice or in male rudiments cultured in the absence of testosterone; (B) male mice or in female rudiments cultured in the presence of testosterone.
C. Response to testosterone of cultured recombinations of ectodermal and mesodermal components of rudiments from normal and *testicular feminization (Tfm)* mutant mice.

with normal mesenchyme, but not when normal epithelium was combined with *Tfm* mesenchyme, indicating that the primary action of the hormone is upon the mesenchymal component, whose response leads in some way to regression and necrosis of the epithelial bud.

Phenotypic sexual characteristics

Hormones are prominent in development wherever changes in widely separated tissues have to be mutually adjusted, sometimes over a long period, as in the control of growth or (especially in embryogenesis at critical periods) as general switch mechanisms where several developing systems must be directed into one pathway or another. The commitment to male or female secondary sexual characteristics, including the development of the ducts of the reproductive system, is a clear example of the latter, determined as a response to androgenic or oestrogenic hormones produced by the testis and ovary respectively. It is much less clear to what extent hormones play a part in primary sex determination, which is in many organisms partly controlled by non-genetic factors; the most extreme example is the marine worm *Bonellia*, whose larvae develop into females if reared in isolation, but into minute semiparasitic males if they become attached to the proboscis of a female. Even in some vertebrates the genetic sex-determining mechanism may be very unstable, as in the fish *Lebistes*, where the genetic sex is often over-ruled by environmental influences. In experimental situations the genetic sex of many vertebrates may be altered by hormonal treatment if it is given sufficiently early in gonadal development. Extensive studies were made by Witschi in the 1930s on the effects of parabiosis (i.e. joining two individuals by grafting at an early stage) in which he showed, e.g. in 1938, that a genetically female toad would be masculinized and produce sperm when grafted to a male, but that there was no effect upon the male partner. Much later, in 1956, Chang and Witschi showed that *Xenopus* males were feminized, to the extent of laying eggs, if oestrogens were added to the water in which they were reared. Among mammals, a condition sometimes arises in the development of cattle as a result of male and female twins sharing a placental circulation, in which the female is characterized by a partial sex reversal of her ovaries and consequently incomplete differentiation of the reproductive tract, and becomes what cattle breeders call a freemartin. This phenomenon was studied in 1917 by Lillie, who concluded that it arose through the effect of androgens passing over from the developing testes of the bull calf and masculinizing the gonads of the female, just as in Witschi's amphibians.

HORMONAL CONTROL OF DEVELOPMENTAL PROCESSES   169

But more recent work, as reported by Short in 1970, suggests this is not the explanation, since it has not proved possible to transform an ovary into a testis or vice versa by hormone treatment in any mammal so far investigated; some circulating substance must be responsible, but its nature and mode of action have not yet been discovered.

The molecular basis of hormone action

The androgens and oestrogens referred to above are steroids—rather small lipid soluble molecules which probably act by passing through the target cell's plasm membrane and acting directly upon the gene regulatory mechanism; O'Malley and colleagues, working on the effects of oestrogen and progesterone on the developing chick oviduct in 1972, produced evidence that these hormones may act at the level of the genome by altering gene transcription, leading to the formation of new nuclear RNA.

The majority of hormones are nonsteroids, and most are polypeptides— much larger than steroids and consequently unable to pass through the plasma membrane, so that their mode of action is quite different. In 1957, working on carbohydrate metabolism in the liver, Sutherland and Rall introduced a concept that has had the most far-reaching influence in research on chemical messengers—that there is a second messenger, cyclic adenosine monophosphate or cAMP (it now appears that another nucleotide, cyclic guanosine monophosphate or cGMP, may be another) which transfers the signal to the metabolic and gene regulatory mechanisms within the cell. Polypeptide hormone molecules are accepted at specific receptor sites which exist only on the target cells, at their cell surface, where cAMP is synthesized through the activity of the enzyme adenylcyclase as a response to the presence of the hormone molecule. Within the cell, cAMP may also be inactivated through the activity of the enzyme phosphodiesterase, related in some way to the hormone molecules at the surface, so the concentration of intracellular cAMP is modulated in both directions. Through its interaction with protein kinases, this has an effect upon the level of many other enzymes, and consequently upon the metabolic state of the cell, which will depend partly upon the cell's previous state; the degree of specificity and flexibility offered by this system is very great, and Bitensky and Gorman in 1972 and McMahon in 1974 have listed a wide range of cellular activities which it has been found to regulate—including morphology, motility and pigmentation of cells, cell division, stability and assembly of microtubules, plasma membrane permeability, gene expression, protein synthesis. Almost all work so far has been upon adult tissues (the closest to a morphogenetic system is in the slime mould, *Dictyostelium*, see chapter 1, where cAMP, secreted outside of the cell, acts as an inducer of cell aggregation) but the cellular activities affected are of the same type as in embryogenesis. We have referred in the last chapter to McMahon's hypothesis that at least some embryonic inductions act through the cAMP system; it is certainly the case that it is established as an extremely widespread and possibly universal mechanism for integrating cellular responses with specific external signals, including hormones and neurotransmitters, in adults; it may well turn out to be just as widespread in embryos, and the distinction between induction and hormonal control will in that case sometimes be difficult to maintain.

Control of metamorphosis in insects

The most striking examples of the deployment of hormones as integrating control mechanisms occur where there is a complete change in an organism's mode of life, requiring drastic changes in the structure and function of many organ systems, i.e. in metamorphosis. This process has been studied intensively in amphibians and insects, and many parallels have been found in the control mechanisms (e.g. see Tata's review of 1971), in which thyroid hormones in the first group and ecdysone in the second play interestingly corresponding key roles. In both, the initial stimulus may be an invironmental signal—length of daylight or a sudden temperature change—transmitted through the nervous system to the brain where, in the hypothalamus in amphibians and the neurosecretory cells in insects, a sequence of hormone signals is set up, which lead to rapid cell division and growth in some systems, and equally rapid necrosis and resorption in others. Here we shall consider the process in insects.

In the more primitive insects (Hemimetabola) metamorphosis proceeds through a succession of stages, the nymphal instars, each one separated from the last by a casting of the old cuticle (ecdysis) to reveal a newly formed cuticle and modification in the size and shape of certain structures, e.g. the wings, towards their adult condition. In this group there are the grasshoppers and locusts (Orthoptera) and plant-feeding and blood-sucking bugs (Hemiptera). In more highly evolved insects (Holometabola)—the ants and bees (Hymenoptera), butterflies and moths (Lepidoptera) and flies (Diptera)—an entirely distinct way of life and an appropriately different form has been evolved for the early instars—the larval stages, which undergoes such a radical transformation into the adult form that a quiescent pupal stage intervenes in which the organism is virtually reconstructed. In both types of metamorphosis—incomplete and complete as they are called—the hormonal control mechanism is essentially the same.

The basis of this mechanism was worked out by Wigglesworth in a series of brilliant studies over many years, beginning in 1934 and reviewed by him in 1976, using as material the blood-feeding hemipteran bug *Rhodnius*, in which there are five nymphal instars (or larval stages) and an adult stage. One peculiarity which makes it particularly suitable for this work is that it takes only one blood meal during each instar and the stimulus for moulting comes from the resulting stretching of the abdominal wall, which produces the neural signal which is acted upon by the neurosecretory cells in the brain. One of Wigglesworth's earliest experiments was to remove the head of a nymph one or two days after feeding, in which case no moult took place; decapitation after this time did not prevent moulting, showing that a critical period was involved. Nymphs from these two experiments could be joined in parabiosis, and in this case both would moult, indicating that some hormonal substance is produced in the head which is circulated through the body.

Wigglesworth also showed that if a 4th stage—or even a 1st, 2nd or 3rd stage—nymph is decapitated very shortly after the critical period it often develops adult characteristics, e.g. in the form of the wings. He showed that a capacity to produce adult characteristics is present in the epidermal cells all through postembryonic life, but that a juvenile hormone is produced in the head region in the first four nymphal stages which inhibits its expression. But if a decapitated 1st-stage nymph is parabiosed to a 5th-stage nymph's head, a tiny adult is produced. By deploying such techniques in a wide variety of ways, he was able to show that in *Rhodnius* postembryonic development is regulated by the changing balance between one hormone which controls growth and moulting, now known as ecdysone, and a juvenile hormone which inhibits the expression of adult characteristics.

Other workers, notably Fukuda, Kühn and Piepho, and Williams and Schneiderman, found essentially the same mechanism operating in holometabolous insects, and progress was made with isolating and characterizing the hormones involved. Both ecdysone and juvenile hormone have also been isolated from a number of plants, in which they probably function as a protection against insects feeding on their sap, by disrupting their hormone balance. Developments in 1964 were reviewed by Schneiderman and Gilbert, and an extensive review which illustrates how wide this research field has become was made by Doane in 1973.

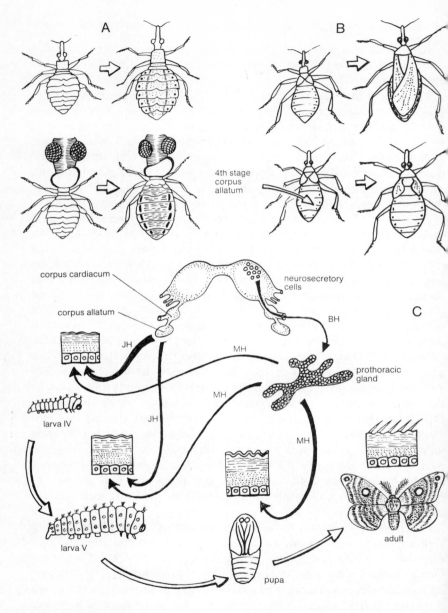

Figure 14.2 Metamorphosis in insects.
The "incomplete" type is shown in the hemipteran *Rhodnius* (A, B) and the "complete" type in the silk-moth *Hyalophora cecropia* (C); hormonal control is the same in both.
A. Transformation of a 1st-stage larva into a 2nd-stage larva in the normal development of

The classical scheme, thus worked out, is as follows, but there are many modifying factors—genetic and environmental—which modify it in particular cases and which led Sláma in 1975 to propose a radical revision of these basic concepts. When the neurosecretory cells are stimulated they secrete a brain hormone (ecdysiotropin) which is passed down the nerve axons to the corpora cardiaca, from which it is liberated into the circulating blood. The brain hormone stimulates prothoracic glands in the thorax to secrete the moulting hormone ecdysone, which is circulated and acts directly upon the epidermal cells, leading to a chain of events—protein synthesis, cytoplasmic growth, mitosis and cuticle deposition—which culminates in ecdysis. A third hormone is secreted by paired endocrine glands attached to the brain, the corpora allata; this is the juvenile hormone which controls the course of the moulting process and the type of cuticle which is formed. In *Rhodnius* and other hemimetabolous insects production of juvenile hormone stops in the last nymphal stage, allowing the epidermal cells to form the adult cuticle. In the moth *Cecropia* and other holometabolous insects juvenile hormone is still secreted even in the last larval stage, but in very reduced concentrations, so that the epidermal cells produce the pupal cuticle; thereafter the level is reduced further and the imaginal cuticle, with all its adult characteristics, is produced, and this involves the eversion of structures such as the eyes, the antennae, the legs and wings, the external genital apparatus, all of which have been developing internally as imaginal discs throughout embryonic and larval life (see chapter 15). Within the pupal cuticle there is a massive breakdown of many of the larval structures, e.g. in *Drosophila* the whole of the abdominal epidermis, and much of the gut and musculature, followed by mopping up by phagocytes and then reconstruction by cells derived from the histoblasts, cells which have existed in small groups (like the imaginal discs) throughout larval life.

Rhodnius (above); joining a decapitated 1st-stage larva to the head of a moulting 5th-stage larva produces a minute "adult" form at the next moult (below).
Based upon V. B. Wigglesworth.

B. Transformation of a 5th-stage larva into an adult in the normal development of *Rhodnius* (above). Implanting the corpus allatum from a 4th-stage larva into the 5th-stage larva produces a "6th-stage" larva at the next moult.
Based upon V. B. Wigglesworth.

C. The classical scheme of the hormonal regulation of metamorphosis in the silkmoth. BH—brain hormone (ecdysiotropin); JH—juvenile hormone; MH—moulting hormone (ecdysone).
Based upon W. W. Doane.

The epidermal cell in insects

The key to much in insect development is the nature of the cuticle and of the epidermal cells which secrete it, and here again Wigglesworth's studies on *Rhodnius*, e.g. in 1961, are of fundamental importance. In its normal state the epidermis consists of a single layer of usually flattened cells, resting on a basal lamina and supporting the cuticle, through which very fine cytoplasmic extensions from the epidermal cells penetrate in pore canals. Under special circumstances these cells are capable of great activity, e.g. when the epidermis is wounded, when neighbouring cells—stimulated by autolytic products arising from the damaged cells—migrate towards and crowd around the edge of the wound, then distribute themselves over the wound itself to restore epidermal continuity, while cell density in the zone which has been left sparsely populated by this emigration is restored through increased mitosis in the remaining cells.

Ecdysone causes the epidermal cell to enlarge and become cuboidal, and to proliferate by mitosis, producing an excess beyond what is required to secrete the new cuticle, so that some die and are phagocytosed by neighbouring cells. These cuboidal cells then go on to produce the new cuticle, which is made up of chitin and protein: first a very thin epicuticle which therefore forms a surface layer which is thrown into characteristic folds (of great use in pattern analysis—see chapter 15); then a thick endocuticle; finally the inner layers of old cuticle are removed by digestion by enzymes secreted by the epidermal cells and the process of tanning (the chemically complex process of hardening) the new cuticle begins.

Evidence of gene activation in insect metamorphosis

In some dipteran flies very prominent salivary glands are formed: in the midge *Chironomus* they secrete a mucoprotein which is used in constructing the cocoon in which the larva lives, and in *Drosophila* they develop towards the end of larval life, when they secrete the protein which glues the pupal case to the substrate. These glands are enormously valuable material for the analysis of genetic regulatory mechanisms, since their cells contain polytene chromosomes, formed by repeated divisions of the chromosomes in which the products of each division remain together so that giant multistranded chromosomes are formed. The opportunity which these chromosomes offer of viewing evidence for gene activation directly was revealed by the work of Becker in 1962 and Clever in 1963; regions can be distinguished in these chromosomes known as *puffs*, produced by synthetic activity of the genes in these regions: the pattern of puffing is changed immediately before pupation, and at any other time if the salivary glands are exposed to ecdysone, reflecting a change in the pattern of RNA synthesis, indicating that the action of ecdysone is upon specifically located genes, and providing strong evidence for control of differentiation at the level of transcription.

CHAPTER FIFTEEN

INSECT DEVELOPMENT

ASPECTS OF EARLY EMBRYOGENESIS IN INSECTS HAVE BEEN CONSIDERED IN chapters 2 and 6, but the most distinctive features of development in this group arise from its continuation in postembryonic stages to the adult stages, separated by the casting of the old cuticle, which is too rigid to permit more than a limited amount of growth within it.

Discontinuities in insect development

This stepwise type of postembryonic development, in combination with extension or contraction of the nymphal or larval period, may also impose a difference in form between adults, leading to polymorphism. The simplest example was described by Huxley in 1932 in the earwig *Forficula*, in which two types of male adult occur, one with short, the other with long, abdominal forceps. The forceps follow a type of differential growth which is widespread in animals, known as *allometric growth*, in which the relative size of two parts of the body change as the total body size increases according to the formula $y = bx^a$, where x and y are the two body parts and a and b are constants, of which a (called the allometry constant) is the one which determines the divergence of the two body parts; in *Forficula* there is sometimes an extra ecdysis in males, producing adults which are not only larger overall, but in which the forceps are disproportionately large. Wilson showed in 1971 that similar discontinuities imposed on an allometric growth pattern account for the much more striking forms of polymorphism which play an important part in the social organization of ants, including the occurrence of soldier castes with relatively enormous heads and jaws (see figure 1.3). In these and other social insects, the proportions of the various castes required vary with circumstances and are controlled ultimately by the circulation of social hormones (pheromones) through the colony.

Another type of discontinuity which is more prominent in insects than in any other group, again a consequence of the evolution of a cuticular exoskeleton since, to allow active movement, it must be developed as a series of rigid plates with flexible connections (as in a suit of armour), is the segmental organization of the body. This is correlated with the existence of switch mechanisms controlling segmental development in different regions of the body which, as we shall see, occasionally break down, so that structures which are appropriate to one segment are actually produced in another, in which they are not. Mechanisms of this sort, controlling the deflection of development into one pathway or another, are basic to all organisms, but their effects are often particularly striking in insects, which therefore constitute such good material for investigations into the problems of determination that this group, from being comparatively neglected, now occupies a place in the forefront of developmental biological studies.

Genetic techniques

Equally important, and even more so as emphasis on studies on the genetic control of development increases, is the range of genetic techniques available in insects, most of all in the fruit fly *Drosophila* in which their sophistication and the wealth of mutants available is immense. Against this advantage is the underdeveloped state in insects of the sort of cell and tissue-culture methods which play such a prominent role in vertebrate embryology, so that studies in insects and vertebrates, though posing the same fundamental problems, tend to be on rather different levels, leaving some uncertainty as to the degree to which results from one group can be applied to the other. A very powerful synthesis of the two may be expected as culture methods improve in the one (as work such as that by Shields and Sang in 1970 on *in vitro* culture of embryonic cells from *Drosophila* suggests that they will) and through increasing use of the mouse (in which a comparable range of mutations and techniques for their manipulation is available as potential tools for embryological research) in the other.

In insect studies mutations are used in two ways: firstly, for altering the course of normal development, particularly of morphogenesis, in the way described for vertebrates in chapter 13. The development of the *Drosophila* wing was studied in this way in 1940 by Waddington, who described the effects of 30 genes on this system, and in 1976 by Whittle, who has reviewed both the advantages and the pitfalls of analysis of this sort. The number of mutations affecting the wing now stands at 222, but this may give a misleading impression of the complexity of genetic activity involved in wing morphogenesis if it is forgotten that many of the effects will be

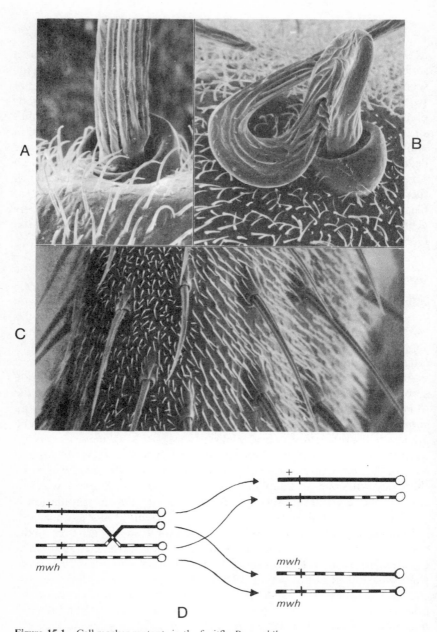

Figure 15.1 Cell marker mutants in the fruitfly *Drosophila*.
A. Portion of a leg showing a bristle (macrochaeta) and hairs (microchaetae). The bristle consists of a long chitinous structure derived from the cytoplasmic extension of a single

essentially rather trivial pleiotropic consequences of genes which play no direct part in controlling wing development. But some mutations enable operations of great precision to be carried out, e.g. by using the technique developed by Russell in 1974 where cell-autonomous temperature-sensitive lethal mutations are used in combination with the technique of somatic crossing-over, described below. Localized clones of the lethal cells are produced at a temperature which prevents the expression of the lethality, then, at any desired stage of development the cells can be killed by transferring the organism to the temperature at which lethality is expressed.

The second use of mutations in insect developmental studies is to provide cell markers which enable individual cells and their progeny to be labelled, and clearly and instantly recognized in a way which is very rare in vertebrate material. The existence of these markers follows from the occurrence in all insects of integumental bristles of one sort or another, formed from one greatly enlarged cell, the *trichogen*. In Lepidoptera the wing scales, visible macroscopically and elaborate in structure, are only modified bristles, still formed from a single trichogen. Bristles arise by the division (in the midpupal stage in the case of adult bristles in holometabolans) of a single cell within the epidermis, which grows rapidly to about 1000 times the size of its neighbouring cells. One of the daughter cells (the *tormogen*) forms a ring which later produces a cuticular socket, through which the other daughter cell (the trichogen) extends a process which, when its cuticle hardens, forms the bristle. Many genes affect the development of the bristles in *Drosophila*, either in its pigmentation (e.g. the yellow *(y)* gene causes the colour to be pale gold rather than dark brown) or in its structure (e.g. the *singed (sn)* gene causes bristles to be stunted and gnarled). In general, the effects of these markers are autonomous to the cells in which they act. They are all expressed in the appearance of the cuticle, and are therefore restricted to the surface layer of cells; but this is of such predominant importance in the structure of insects that it is not a severe limitation.

(trichogen) cell, and is supported at its base in a socket formed from another (tormogen) cell. Hairs are likewise developed from single cells.
B. The gnarled bristle of the *singed (sn)* mutant.
C. A genetic mosaic in which a clone of cells homozygous for the recessive mutant gene *multiple wing hairs (mwh)*, in which each cell produces several hairs, has been produced by somatic crossing over in a heterozygote at a mitotic division within the developing disc, as illustrated in D.
Stereo electron-micrographs by courtesy of P. J. Bryant.

In order to make use of such markers they must be introduced into the developing organism, and at a known time. This is achieved by the creation of genetic mosaics, in which some of the cells are of the marker genotype. This is most commonly done by inducing chromosomal crossing-over, and consequently genetic recombination, in somatic (i.e. non-germline) cells at a selected stage by X-irradiation. Matings are arranged so that the genotype of the fertilized egg carries recessive marker genes in heterozygous condition, so that normally they would have no effect upon the phenotype. Irradiation may cause somatic crossing-over to occur in a few isolated cells, in which one or other of the marker genes may be brought into homozygous condition in the daughter cells, so that the progeny of such a cell—the clone—arising from it will be recognizable through the phenotypic effect of the mutant gene. For example, the original genotype might be $\frac{y\,sn^+}{y^+\,sn}$, producing a wild-type fly, but mitotic recombination might give daughter cells homozygous for *yellow* $\left(\frac{y\,sn^+}{y\,sn^+}\right)$ and *singed* $\left(\frac{y^+\,sn}{y^+\,sn}\right)$ respectively. Many aspects of development can be studied by the use of such clones, e.g. since the size of a patch produced by such a clone of marked cells depends upon the number of cells which have arisen through mitosis since the time of irradiation, information can be obtained about the numbers of cells present at that time and the subsequent rates of cell division; Bryant and Schneiderman used this method in 1969 to show that the adult *Drosphila* leg is formed from a group of about 20 cells arising in the embryonic blastoderm, which grows throughout larval life with an average cell cycle time of about 15 hours.

Another method of producing genetic mosaics was used by Hotta and Benzer in 1972 to construct fate maps for the early *Drosophila* embryo with a view to identifying the structures in which various genes affecting behaviour exerted their primary effects—whether within the affected organ or remote from it, e.g. in the central nervous system. It is impossible to use the simple vital staining method used in vertebrate embryos, and the technique actually employed involved the production of gynandromorphs (i.e. a particular type of genetic mosaicism in which male and female cells occur together). A stock was used in which one of the X-chromosomes of the female embryo is generally eliminated from one of the two lines of cells established by the first division of the zygote nucleus, producing an embryo in which approximately half the cells are genetically male and half female; but since the orientation of the division is at random, the boundary line may occur anywhere on the surface of the resulting blastoderm and, subsequently, of the adult fly in which the marker genes are phenotypically expressed. The frequency with which the boundary line falls between two regions of the blastoderm is therefore related to the distance between them and, by a quantitative analysis of which adult structures are male and which female, their distance apart and, by triangulation, the position of the corresponding ancestral cells in the blastoderm can be calculated. In using the technique for locating sites of behavioural lesions, behaviour genes are combined with cuticular markers; if the disturbed activity and the marker phenotypes were always exhibited in the same adult structure (e.g. a leg tremor with marked legs) it would suggest that the primary lesion occurred within the leg itself, say in the musculature, but if this is sometimes not the case the site of the primary lesion must be elsewhere, so that the mosaic boundary line has passed between the two separate blastodermal regions concerned, in which case its location can be calculated as described. Thus for the *Hyperkinetic* gene *Hk*, causing leg shaking under ether anaesthesia, the data indicate that the primary lesion is in the ventral nervous system, and neurophysiological findings support this conclusion.

Cuticular patterns in hemimetabolous insects

Investigations on the blood-sucking bug *Rhodnius* and the milkweed bug *Oncopeltus* have provided important information about the control of antero-posterior polarity in the epidermal cells of the abdominal segments in other insects. This polarity is exhibited in the cuticle they produce, in *Oncopeltus* directly through small hairs or bristles which point posteriorly.

In 1966 Lawrence investigated the gaps which occasionally appear between segmented boundaries in *Oncopeltus*, and described an interesting pattern of bristles surrounding them. A circular area is defined, with the gap forming its transverse diameter, within which A-P polarity is reversed, except towards the periphery, where there is a transition to the normal orientation. Lawrence proposed that the normal segmental polarity depended upon a gradient set up with a high point at the anterior segmental boundary and a low point at the posterior boundary, and proposed a simple model to illustrate some of its features. In this the gradient is formed of sand, supported at its high point by a barrier, arrows indicating polarity always pointing downhill. If a gap is made in the barrier, the sand flows through it and arrows on the sand pointing down the resulting slopes show exactly the pattern of bristle orientation in this situation.

Another way of indicating the pattern would be to draw contour lines around the slopes joining points of equal height, and in *Rhodnius* the

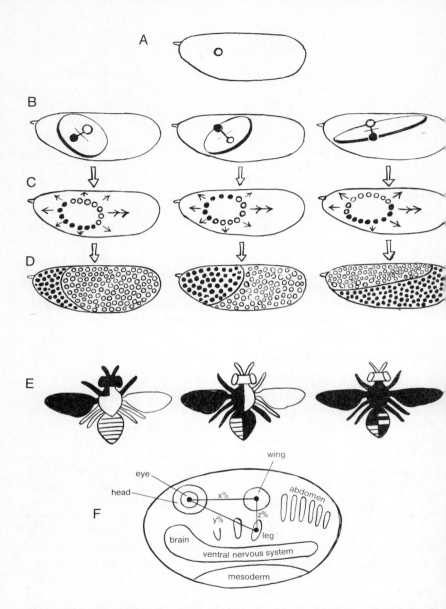

Figure 15.2 Construction of a presumptive fate-map of the *Drosophila* blastoderm, using the genetic mosaic technique.
A. Egg with undivided zygote nucleus (XX), in a female embryo.
B. The zygote nucleus divides with the mitotic spindle orientated at random, the plane of division indicated here as a disc passing at right angles through the spindle. At this division the X-chromosome is eliminated in the black nucleus (XO).

polarity of the epidermal cells is indicated by uninterrupted cuticular ridges running transversely across the segment, at right angles to their antero-posterior axis, constituting virtually natural contour lines on the proposed gradient. In 1959 Locke had made interesting experiments on *Rhodnius* cuticle patterns which were repeated and reinterpreted by Lawrence, Crick and Munro in 1972 in the light of positional information theory.

Locke's experiments, in which he transplanted squares of dorsal segment in 5th-stage nymphs and analyzed the resulting cuticular patterns in adults are illustrated in figure 15.3. Squares were grafted to the same position in hosts, with or without rotation through 90° or 180°, and transplanted without rotation from an anterior to a posterior position within the same segment. The last is crucial in establishing that polarity is due to a recognition of the cell's position in the segment, rather than depending on some inherent polarity within the cell, in which case the pattern would be unaltered. All the resulting patterns conform to the sand-gradient model.

In this model there are two opposing forces maintaining the slopes, corresponding to the balance between the effects of gravity creating a new orientation and friction conserving the old one. In the organism, the first corresponds in Lawrence, Crick and Munro's more elaborate model to a diffusion gradient of some morphogen providing positional information, set up as suggested, and the conservative force is created through each cell attempting actively to maintain its own "remembered" concentration of morphogen. When a cell finds itself in a new position, it reaches an equilibrium somewhere between its former level and the concentration around it at its new level. The hypothesis that cells migrated after transplantation was rejected, but it was subsequently shown in 1974 by Lawrence himself (and independently by other workers) that rotation of a graft back towards its original orientation does in fact occur, suggesting that cells do carry a remembered value, and that they move round until this value matches that of the surrounding host cells. Similar considerations in experiments on the development of scale patterns in lepidopteran wings led Nardi and Kafatos in 1976 to suggest an interesting alternative to diffusion

---

C. The sphere of dividing nuclei expands (see figure 2.1) and (D) the nuclei populate the egg cortical cytoplasm to give the blastoderm, half of which consists of a clone of female (XX) and half of a clone of male (XO) cells.
E. Recessive marker genes on the X-chromosome which has not been eliminated are expressed (black areas) in cuticular structures in the adult fly.
F. The distance apart (frequency boundary falls between) and the position of presumptive regions on the blastoderm can then be calculated (see text). The presumptive nervous tissue and presumptive mesoderm have been established from descriptive studies. Based upon Y. Hotta and S. Benzer.

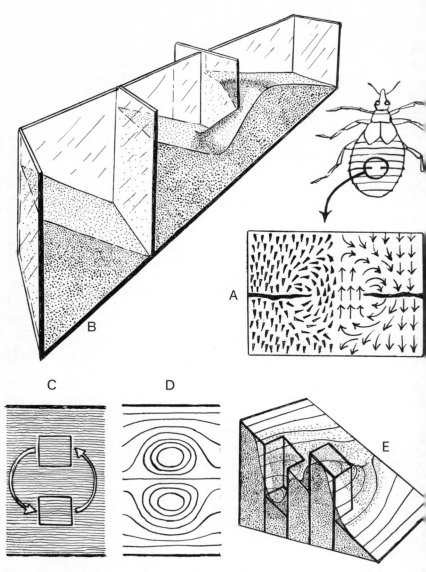

Figure 15.3 Cuticular patterns in abdominal segments of the hemipterans *Oncopeltus* and *Rhodnius*.
A. Pattern of bristle orientation in an *Oncopeltus* larva where a break has occurred in the boundary separating one segment from another.
B. Polarity of the bristles may depend upon a gradient of positional information, represented in this model as gradients of sand, set up in a series of glass compartments whose walls represent the segmented divisions of the abdomen; the model is shown sectioned down the

models in the form of a gradient of cell adhesiveness, based upon Steinberg's hypothesis of cell sorting through differential cell adhesion; but if this is accepted, the problem of setting up the gradient in the first place still remains.

Clonal analysis of segmentation in *Oncopeltus*

Whatever the nature of the gradient, it is repeated in all abdominal segments. Further experiments on *Oncopeltus*, carried out by Lawrence in 1973, explored the nature of the segmentation process using a genetic marking method to identify cell clones. In this case the somatic crossing-over technique is not available and, instead, somatic mutations were induced in some of the epidermal cells by X-irradiation of eggs and larvae, producing clones of cells coloured in a variety of ways—red, dark-orange, white—other than the normal orange-yellow; analysis of these clones in the 5th nymphal stage gives information about the way in which segments develop and are demarcated. It is known that each segment is produced from four groups of embryonic cells, each of which produces a segment quadrant, a right and left dorsal quadrant, and a right and left ventral quadrant; the experiments concern the demarcation of these segmental quadrants in development. No matter how early the irradiation—even 6–12 hours after egg-laying when nuclei are still moving to the cortical cytoplasm—no clonal patch ever comes to fill a whole segment quadrant in the 5th-stage nymph, indicating that some cell mixing occurs during cleavage, and that the primordium of each segment quadrant consists of a group of cells which is not a clone. Some clones produced by irradiation at this time are found to extend over more than one segment, indicating that the segmental boundaries are not established by 12 hours.

Clones initiated at any stage are often broken up into patches, showing that cell mixing continues to occur even after the formation of the cellular blastoderm which is formed between 12 and 20 hours. Irradiation during

---

mid-line. If there is a slit in the glass between two compartments, and the halves are pulled out to give a gap, the sand falls into new gradients which give the observed patterns of polarity at the gap.
Based upon P. A. Lawrence.
C. Patterns of ripples in the larval cuticle of an abdominal segment in *Rhodnius*. Transplantation of squares of epidermis and overlying cuticle as shown produces, at the next moult, the pattern of ripples illustrated in D. This may be explained by reference to the sand model illustrated in E, where the sand gradient represents a gradient of positional information on which ripple development is based.
Based upon M. Locke and P. A. Lawrence.

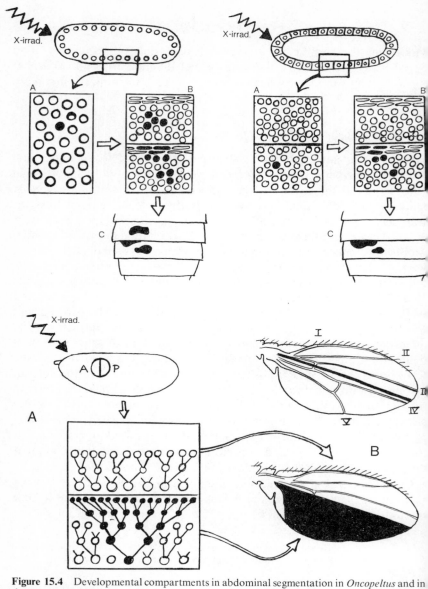

Figure 15.4 Developmental compartments in abdominal segmentation in *Oncopeltus* and in the *Drosophila* wing.
Top
X-irradiation of the *Oncopeltus* embryo at the blastoderm stage causes somatic mutation in an occasional nucleus (black), which by division produces a clone of cells carrying the mutated gene, expressed as a pigment difference in the 5th-stage larval cuticle. The early division products of the clone become dispersed among neighbouring cells, leading to the development

this period produces an increasing proportion of clones which are confined to one segment, and after 20 hours all are of this type. Where clonal patches occur at a segmental border in the 5th nymph, they terminate in a straight line, so that the cell mixing which does occur must always stop short at the boundary. This indicates that, coinciding with the period of cellular formation in the previously syncytial blastoderm, some process of demarcation occurs which prevents a cell lineage within the segmental quadrant primordium mingling with those within adjoining segments. In the nymphal stages, the anterior margin of each segment consists of spindle-shaped cells, and mitoses are orientated accordingly, but in the earliest embryonic stages there is no histological feature marking the boundary; the mechanism of demarcation is unknown, but it must presumably involve the establishment of distinctive cell surface properties. Segments can be seen soon after the formation of the germ band, but the clonal analysis shows that the segments have independent cell lineages before this occurs. Clone patches produced by irradiation at 13 hours show that each patch covers about 10% of the segment quadrant, indicating that at the time when the ancestral cell mutated there were 9 others in the primordium which did not, and that each group consists therefore of about 10 cells. Following the work of Garcia-Bellido (p. 192), the segment quadrant is recognized as a developmental compartment whose cells form a polyclone (i.e. all of the descendents of a small group of cells, which is not a clone, set aside early in embryonic development) and it illustrates in a simple way a principle of development which is certainly of great importance in insects and whose significance may extend to other organisms.

---

of several pigment patches. If *(left)* the mutation occurs during the syncytial blastoderm stage, the patches may be found in more than one segment, but if *(right)* mutation occurs during the cellular blastoderm stage, pigment patches are confined to one segment, indicating that cells at this stage restrained by some sort of compartmental boundary.
A. Nuclei or cells at time of irradiation.
B. Arrangment of epidermal cells in abdominal segments in larval stages.
C. Mutant pigment patches in 5th-stage larval stages.
Based upon P. A. Lawrence.
Bottom:
A. X-irradiation of the *Drosophila* embryo at the cellular blastoderm stage, when cells of the presumptive limb area consist of a polyclone divided into anterior and posterior compartments. The compartmental boundary in the adult wing (B) is demonstrated by producing a marked clone, by somatic crossing over, in which the cell proliferation rate is much greater than in cells of original genotype.
Based upon A. Garcia-Bellido, P. Ripoli and G. Morata.

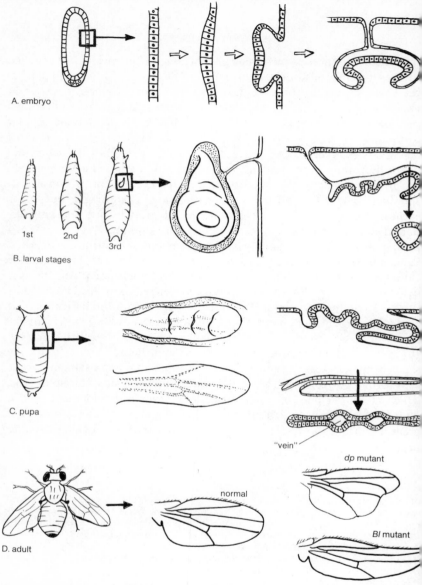

Figure 15.5 Imaginal discs and the development of the wing in *Drosophila*.
A. Formation of the wing disc in embryogenesis.
B. Development of the wing disc in the three larval stages.
C. Appearance of the wing "veins" and eversion of the disc in the pupal stage.
D. The adult fly showing the fully expanded and hardened wing, and *(right)* two wing mutants, *dumpy (dp)* and *Blade (Bl)*.
Based partly upon P. J. Bryant.

Imaginal discs in holometabolous insects

The dramatic change from the larval to the adult condition which characterizes holometabolous insects does not entail that all the adult structures arise afresh in the intervening pupal stage. On the contrary, all of the integumentary cuticularized parts of the adult are formed from primordia which are set aside in the embryo and which grow within the body cavity throughout the larval period. These primordia are the imaginal discs, whose existence has been known since the eighteenth century, when a fascination with the extraordinary transformations of insects misled many fine naturalists into the wildest extremities of preformationist theory (see chapter 1), but no adequate study of their developing structure was made until that of Poodry and Schneiderman on the imaginal disc of the leg of *Drosophila* in 1970.

Imaginal discs are essentially hollow flattened sacs of epidermal cells, formed by invagination of the embryonic epidermis and retaining a connection with the larval epidermis by a very attenuated tubular stalk of cells. The bottom of the sac becomes folded back into the sac and folded in ways which are characteristic of the various discs, of which there are seven major pairs and one fused pair in *Drosophila* eye/antennal, 1st, 2nd and 3rd thoracic legs, prethoracic dorsal discs, wings, halteres (the small knob-like wing vestiges of dipterans) and the single genital disc.

At metamorphosis, eversion of the discs occurs, when through changes of cell shape, each emerges through its stalk and expands to give the characteristic adult form, secreting first a pupal cuticle and then an adult cuticle which, after the insect has broken through the pupal case, expands again and hardens—most dramatically and familiarly in the butterfly's wing—to give the final adult structure. Each major disc is made up of only between 15 and 60 cells when it first appears in the embryo, but grows rapidly within the larva and contains many thousands of cells at metamorphosis; but apart from the characteristic folding, no differentiation occurs until the pupal stage and, unlike the epidermal cells, they secrete no cuticle until then. Yet, from its first appearance, each disc is determined to form one adult structure or another, and this long interval between determination and differentiation gives the imaginal disc a particular interest.

Transdetermination in *Drosophila* imaginal discs

In 1966 Hadorn summarized his remarkable studies in which he showed that this interval between determination and differentiation could be

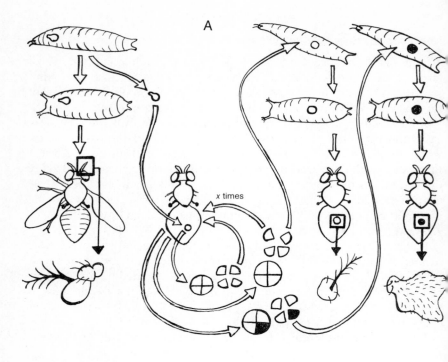

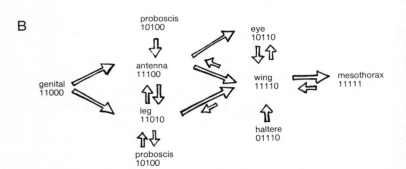

Figure 15.6 Transdetermination in *Drosophila* imaginal discs.
A. The technique of maintaining cultures of imaginal disc cells within the abdomens of adult flies. In the case illustrated, fragments of antennal disc continue to produce antennal structures when tested *(white)* but after repeated transplantations transdetermination has occurred in one fragment *(black)* to produce wing structures when tested. Based upon E. Hadron.

extended virtually indefinitely by bypassing the hormonal situation which causes the cells to differentiate, by taking imaginal discs from full-grown larvae and implanting them into the abdominal cavities of adult flies, in which the discs would continue to grow without differentiating. In order to bring about differentiation, it was necessary merely to transplant fragments of disc back into a metamorphosing larva; the resulting fragments of adult structures can be recognized as parts of particular adult organs. In general, the original determined state of disc fragments persists for years of culturing *in vivo* in this way, i.e. over hundreds of cell generations, showing how extremely stable it is; but occasionally—after a number of transplantations—the determined state of some cells becomes changed, resulting in differentiation as parts of a disc of some other sort, so that a fragment derived from a culture of genital disc cells may differentiate instead as part of an antenna or a leg. Such a change of a cell's determined state Hadorn called *transdetermination*. In 1967 Gehring showed, using the somatic crossover marker method, that it occurs in groups of cells rather than in a single cell. Cells derived from such fragments continue in culture in the same transdetermined state, but may once again undergo a further transdetermination, all these changes occurring with characteristic probabilities and only in particular directions for which a flow diagram can be constructed, e.g. genital disc cannot transdetermine directly to wing, but must first change to antenna or eye, either of which can transdetermine to wing. On the basis of these facts, Kauffman in 1973 proposed an interesting model for the control of determination in imaginal discs based upon five circuits, each with two stationary states, each disc being characterized by the combination of states selected by the different circuits. This model, and the experimental data which support it, illustrate the potentialities of this type of material for giving flesh to the bones of determinative theories of the sort described in chapter 12.

Regulation in imaginal discs

Though each disc is committed to a particular pathway of development from its first appearance in the embryo, the cells of which it is composed are not committed to develop as one part of the disc rather than another; the

---

B. Flow diagram of transdetermination state changes which may occur in imaginal discs of *Drosophila*, showing how it may be based upon switches in five independent circuits, each with two stationary states (0, 1).
Based upon S. A. Kauffman.

disc is in fact an embryonic field within which regulation is possible. In 1974 Bryant showed that, if a disc is divided into two approximately equal parts, one of the halves generally regenerates to form the complete structure at metamorphosis, while the other half produces a mirror-image duplication of the corresponding half-structure. The hypothesis that there are polar co-ordinates of positional information within the field, proposed by French, Bryant and Bryant in 1976 and referred to in another connection— vertebrate limb regeneration—in chapter 12, convincingly explains how this may come about. In the *Drosophila* leg disc, for example, radial co-ordinates originate at the distal tip of the disc, and circular co-ordinates encircle it; the outermost circle is at the proximal disc boundary and may be visualized as a clock face, with positional values 1 to 12. As in the vertebrate limb, growth occurs to intercalate arising positional values always by the shortest of the two possible routes; consequently, when a disc is divided and its edges heal together, one "half" of the disc will include slightly more than half the clock face (say 0–7) and regenerate the rest (8, 9, 10, 11), while the other will include slightly less than half (say 7–12) and, since intercalation must be by the shortest route (8, 9, 10, 11), producing a mirror-image structure. Not only does the model explain the type of patterns formed (though not as yet in molecular terms) but, since growth is stimulated when two non-contiguous positional values are brought together and ceases when intercalation of missing values is complete, it also provides a working hypothesis for the study of growth control.

Compartments and homeotic genes in *Drosophila* development

The subject of compartments has been introduced by reference to the establishment of segmental boundaries in *Oncopeltus* where, from an early stage in embryogenesis, cells of two adjoining compartments lie together, and are even connected by gap junctions, yet do not mingle or cross the invisible boundary between them. The concept was formulated in 1973 by Garcia-Bellido, Ripoli and Morata in a study of the developing wing disc of *Drosophila*, and general aspects of the subject were reviewed by Crick and Lawrence in 1975, and by Lawrence and Morata in 1976. In the *Drosophila* wing disc the particular interest of the subject is more obvious since, whereas in *Oncopeltus* the boundary corresponds to the morphological division which appears later between segments, in the wing there seems to be no such morphological or functional distinction between structures derived from adjoining compartments, and the problems of how and why development should entail a mechanism for demarcation of areas which are indistinguishable without sophisticated cell-marking techniques become intriguing.

The somatic crossing-over technique was used, with a variety of recessive mutations such as *singed (sn)*. Irradiation at any larval stage produces clonal patches in the adult wing; irradiation at early stages gives large patches, but infrequently, since there are few target cells in the disc, whereas irradiation at late stages, though it gives many more, produces very small clones, since there are fewer cell divisions to follow. Under these circumstances, the recognition of a compartment boundary is difficult, especially where there is no morphological boundary to serve as a guideline, but the difficulty is overcome by using in addition to the marker genes the dominant mutation *Minute (M)* which, in the heterozygous condition in which the stock was prepared, causes very slow growth, with a low mitotic rate. Recombination after irradiation gives some M^+/M^+ cells in which normal growth is restored and which consequently produce large clones which might be expected to overgrow the normal boundary. But they do not; instead they produce a clonal patch which spreads in a straight line along the boundary between compartments, most strikingly in the boundary which lies close to but not on the IVth wing "vein", and which separates anterior from posterior dorsal wing compartments. Wieschaus and Gehring showed in 1976 that the wing disc and leg disc probably originate together as a polyclone in the embryonic blastoderm and become separated as separate groups of cells a few hours later; the original polyclone becomes divided into anterior and posterior compartments, so that from an early stage four polyclones exist—anterior and posterior parts of the thorax, and anterior and posterior parts of the wing; later, in the first larval stage, dorsal and ventral compartments are separated in each of the polyclones and more subdivisions probably follow later.

Garcia-Bellido has proposed that the polyclones and the compartments they generate are units for the genetic control of development and that, when a polyclone divides, a special selector gene is switched on in one of the daughter polyclones and off in the other, committing the cells to different developmental programmes. Since several subdivisions are made in the course of development, each compartment is specified by a unique combination of active and non-active selector genes; it may be this combination which constitutes the state of determination of its constituent cells.

The subject of compartments is closely related to that of homeotic genes, which are common in *Drosophila* and lead to the substitution of a structure normally formed in one segment for another which is normally formed in

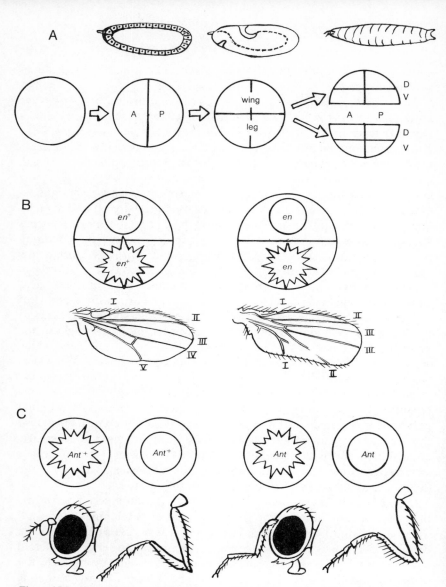

Figure 15.7 Compartments, selector genes and homeotic mutants in *Drosophila*.
A. Embryonic and larval stages, with corresponding sequence of compartmentalizations in the polyclone of cells from which the wing and leg develop.
B. Gene-activation at the *engrailed* locus in the posterior compartment of the developing wing, producing normal posterior wing structures where the wild-type allele is present *(left)* and partial reduplication of the anterior wing structures when the *engrailed* mutant gene is present in homozygous condition *(right)*. In the anterior compartment the gene is inactive.

another segment (e.g. the formation of a leg instead of an antenna in *Antennapedia*) or to a similar substitution of parts within the same segment. It is probably a general rule that homeotic genes, acting in polyclones as the units of determination, transform whole compartments by their effect upon the activity of the selector genes. Thus the mutant *engrailed* produces a mirror image duplication of the anterior part of the wing. Whether the engrailed gene or its normal wildtype allele is present in cells of the anterior compartment of the wing is evidently irrelevant, since the gene is switched off there; but in the posterior compartment normal development depends upon the presence of the wildtype allele and, if engrailed is present instead, the cells differentiate according to the programme for the anterior compartment. In *Drosophila* certainly (and presumably in all insects) the hierarchical subdivision of the developing organism into immiscible groups of cells, whose development is autonomously committed to one pathway or another by the combinatorial state of a few selector genes, constitutes a very important aspect of developmental control. Whether this is so in other organisms, especially vertebrates, remains to be seen.

C. Gene-activation at the *Antennapedia* locus in the antennal compartment, producing a normal antenna where the wildtype allele is present in homozygous condition *(left)* and the development of a leg in the antennal site where the homeotic mutant gene *Antennapedia* is present *(right)*. In the leg compartment the gene is inactive.
Based upon A. Garcia-Bellido.

CHAPTER SIXTEEN

MAMMALIAN DEVELOPMENT

THIS SUBJECT IS NATURALLY OF ESPECIAL INTEREST IN THAT OUR ABILITY to prevent and alleviate congenital abnormalities in man—inherited defects and the effects of pathogenic organisms, deleterious chemicals and other factors in the uterine environment—will depend upon advances in this field, which because of the inaccessibility of the embryo within the mother has been a particularly difficult one for investigation by the traditional techniques of experimental embryology. But in addition, making it increasingly attractive to developmental biologists, is what also gives insects their particular importance—the wealth of genetic information and material, especially in man (where it is chiefly clinical) and in the mouse, available for use in different ways in these two species for research into developmental problems. Because of its special importance in this respect we shall concentrate here on embryogenesis in the mouse.

Special features of amniote embryos: extraembryonic membranes

Amniote animals (reptiles, birds and mammals) differ from anamniotes (fish and amphibians) in that fertilization of the oocyte takes place within the oviduct, and in that thereafter embryonic development takes place either in a terrestrial environment inside a thick-shelled egg or, parasitically, within the body of the mother. Extraembryonic membranes have been developed in reptiles and birds to enable embryos to survive in the latter—the yolk sac, amnion, chorion and allantois, the last two of which have been elaborately modified in mammals to produce, in conjunction with tissues of the uterine wall, the placenta which serves as an organ of physiological exchange between mother and foetus. The amnion exists in all these groups as a sac containing amniotic fluid within which the embryo is protected and supported.

The development of these membranes is most easily explained for the chick embryo where the extra-embryonic somatopleure (ectoderm with its

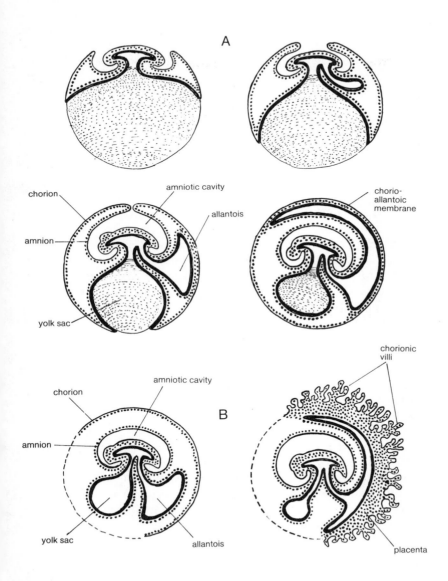

Figure 16.1 Development of the extra-embryonic membranes in birds (A) and their modification in mammals (B).

underlying layer of somatic mesoderm) of the blastoderm folds up and over the embryo all around it, fusing and then dissolving where the folds meet to form an internal membrane (the amnion) and an external one (the chorion). The extraembryonic splanchnopleure (endoderm with its overlying splanchnic mesoderm) gradually spreads over and encloses and absorbs the yolk as blastoderm epiboly proceeds, forming the yolk sac. Meanwhile, a diverticulation from the hindgut rudiment grows out into the extraembryonic coelom, forming at first merely a depository for nitrogenous excretory products; this is an outpushing of the gut endoderm, with its overlying covering of mesoderm. As development continues, the allantois grows enormously, presses against the chorion, which itself becomes pressed hard up against the shell membranes, and the mesodermal layers of these two layers fuse and become highly vascularized to form the embryonic respiratory surface.

In the course of mammalian evolution the developmental origin of the extraembryonic membranes has been so much modified that it is impossible to describe in such a simple way; but the end result is that an amnion is formed which serves the same function as in birds, a small but vascularized yolk sac is formed—though in the mammalian embryo there is virtually no yolk—and a chorion and allantois are formed and contribute to the placenta, going beyond their respiratory role to function in nutritional and excretory exchange as well. The embryo, enclosed in its membranes, sinks into the uterine wall in the process of implantation, and vascular villi grow out into it from the chorion; the placenta is produced by the very close association of the extraembryonic and the material components, brought about by breakdown of tissues which separate the foetal from the maternal vascular systems. In the mouse all intervening tissues—on the maternal side, the epithelium uterus wall, the connective tissue and even the endothelial wall of the blood vessels; and on the foetal side the chorionic epithelium and the connective tissue—are eroded with the exception of the endothelial wall of the foetal blood vessels; the latter is preserved in all types of placentae, providing a barrier against mingling of the two blood circulations, which would—and does when it occasionally occurs—have disastrous immunological consequences for the foetus, since it would be rejected as antigenically foreign. Details of placental structure vary enormously from one group of mammals to another.

Embryogenesis in the mouse: preimplantation stages

A very full description of mouse development was provided by Rugh in 1967, and accounts of early stages by Snell and Stevens in 1966, and Bellairs in 1971. The egg is enclosed in a zona pellucida—a thin membrane secreted by the follicle cells—and, since it contains almost no yolk, cleavage is unimpeded and divides the egg into 2, 4, 8 ... blastomeres and so on until a solid ball of cells (the morula) is formed, quite unlike the blastoderm of the avian egg. The early stages of development are correspondingly different up to the stage of gastrulation, when a primitive streak appears, and from neurulation onwards embryonic development is essentially the same as in birds.

While cleavage is going on the embryo moves down the oviduct, and at the 32 or 64-cell stage is in the uterus. At this stage a fluid-filled blastocoele cavity appears asymmetrically, converting the morula into a blastocyst, differentiated into an external trophectoderm layer and an inner cell mass. The cells of these two structures have distinct properties; those of the trophectoderm are bound together by tight junctions, which allow them to withstand the fluid pressure, and will not aggregate in culture, whereas cells of the inner cell mass do so readily. In 1957 Dalcq had proposed on the basis of histological observations that development of the mouse embryo was mosaic, and that this differentiation occurred through differential distribution of cytoplasmic determinants in the egg. An alternative explanation is that their fate is a function of their position—that those cells with some exposure to the outside environment become determined as trophectoderm cells, and those which are surrounded on all sides by other cells become determined as inner cell mass. In 1972 Hillman, Sherman and Graham showed this to be the case by labelling embryos at the 4- and 8-cell stage with tritiated thymidine, disaggregating the blastomeres and sticking them on the outside of unlabelled 4- to 16-cell embryos, giving them an additional layer consisting of labelled cells. Subsequently autoradiographic analysis at the blastocyst stage showed that most, and in some cases all, of the labelled cells appeared in the trophectoderm, though, if their fate had been already determined at the earlier stage, some of them should have differentiated as inner cell mass.

At this stage, at 4 days, the blastocyst begins pulsatory movements—demonstrated in cine studies by Cole in 1967—which lead to it breaking out of the zona pellucida, ready for implantation if the hormonal conditions in the uterus be right. The blastocyst adheres to the uterine epithelium by the trophectoderm, which penetrates into the wall and is thereafter known as the *trophoblast*, so that the embryo comes to lie within a crypt in the uterine mucosa where it begins its parasite-like association with the maternal tissues.

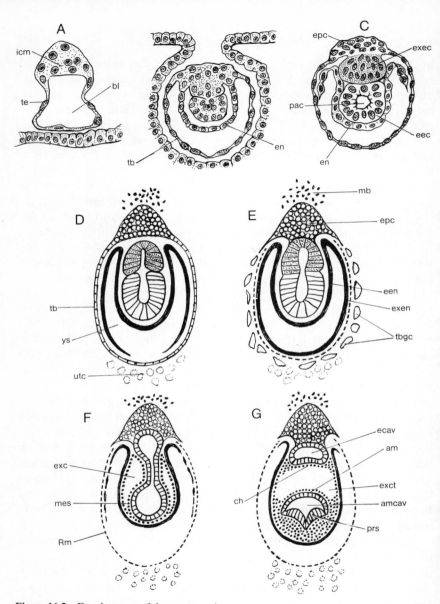

Figure 16.2 Development of the mouse embryo.
Blastocyst at uterine epithelium (A); invading uterine wall (B); and at 5–6 days, shortly after implantation (C).
D–G. Diagrams of embryonic development from 5½ to 7½ days.
Based upon G. D. Snell and L. C. Stevens, and R. Rugh.
For the earliest stages of mouse development see figure 3.2.

Postimplantation stages

Gastrulation begins as implantation occurs, but in a way which appears unlike the events described in chapter 6, depending upon delamination of cell layers rather than on morphogenetic movements, but as soon as the primitive streak is formed the process becomes familiar again. The inner cell mass elongates and is now called the *egg cylinder*; its outer part becomes split off as a separate cell layer which forms the endoderm; its core forms the ectoderm, clearly divided by staining properties into extraembryonic ectoderm nearest the trophoblast, and embryonic ectoderm nearest the blastocoele cavity. Over the embryonic ectoderm the endoderm will form chiefly the embryonic gut wall, but elsewhere this "proximal" endoderm is extraembryonic, and from the base of the cylinder this layer extends as a "distal" layer as a lining to the trophoblast; the blastocoele—now called the *yolk sac*—is therefore becoming enclosed by extraembryonic endoderm which forms the walls of the yolk sac.

The trophoblast adjacent to the inner cell mass gives rise by proliferation to the ectoplacental cone, which invades the maternal tissues as it grows, rupturing blood vessels in its path. Spaces develop within it which become filled with circulating maternal blood, and later it is invaded by blood vessels of the allantois to produce the placenta. The cells in the remainder of the trophoblast undergo no further division but transform into giant cells whose function is unknown. Before this occurs a thin membrane appears between it and the outer yolk sac wall—Reichert's membrane, peculiar to rodents—which consists of extracellular material and which forms a third membrane (with the amnion and the yolk sac) protecting the embryo.

At $5\frac{1}{2}$ days a proamniotic cavity appears in the centre of the embryonic ectoderm, and another cavity in the extraembryonic ectoderm; the two become joined by a lumen and form the amniotic and the ectoplacental cavities respectively. The egg cylinder expands to fill the yolk cavity, and

---

Key:
 am — amnion
 amcav — amniotic cavity
 bl — blastocoele
 ch — chorion
 ecav — ectoplacental cavity
 eec — embryonic ectoderm
 een — embryonic endoderm
 en — endoderm
 epc — ectoplacental cone
 exc — exocoelom
 exec — extraembryonic ectoderm

 exen — extraembryonic endoderm
 icm — inner cell mass
 mb — maternal blood
 mes — mesoderm
 pac — proamniotic cavity
 prs — primitive streak
 Rm — Reichert's membrane
 tb — trophoblast
 tbgc — trophoblast giant cells
 te — trophectoderm
 utc — uterine cells
 ys — yolk sac

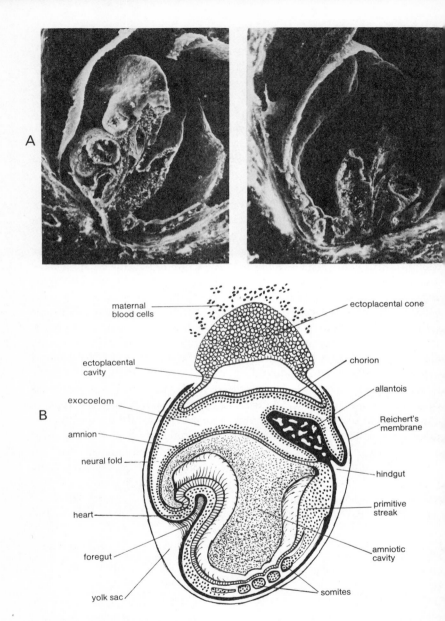

Figure 16.3 The mouse embryo at 8 days.
A. Stereoscan electron-micrograph of an embryo which has been cut in half along the midline, but slightly at an angle so that one half *(right)* contains all of the tail and the other half *(left)* most of the head. The whole embryo can be mentally reconstructed by imagining turning the right-hand figure over onto the left.

the outer yolk sac wall subsequently breaks down, leaving the inner wall to press against Reichert's membrane and through that to establish close contact with the uterine tissues; the effect is as if the yolk sac had been turned inside out.

The relation of the germ layers in the embryonic region is now the reverse of that generally found in vertebrates, with a thick inner ectodermal layer and a thinner outer endodermal layer. The primitive streak appears at this point, beginning as a thickening of a local area at the margin of the ectoderm (which should strictly be called *epiblast* since presumably, as in the chick, mesoderm as well as ectoderm cells are derived from it), followed by the appearance of mesoderm cells moving out between ectoderm and endoderm and penetrating into the extraembryonic region, where the mesoderm splits to give an extraembryonic coelom (exocoelom). As the exocoelom expands, the mesoderm lining it becomes associated with "embryonic" ectoderm on one side, giving the amnion (which strictly makes that ectoderm nonembryonic), and with the extraembryonic ectoderm on the other, giving the chorion. Thereafter, neurulation and somitogenesis proceed just as in the avian embryo, only with the dorsal side of the embryo arched into a U-shape, with the tail almost touching the head; gradually this is reversed, so that the general C-shaped vertebrate embryo form is produced. Between $8\frac{1}{2}$ and 9 days organogenesis is greatly accelerated (e.g. the forelimb bud appears) and exposure to teratogens and other environmental disturbances during this period is especially likely to produce developmental abnormalities. The allantois grows out into the exocoelom and fills it, fuses with the chorion and joins the embryo to the ectoplacental cone, forming the rudiment of the umbilical cord.

Genes and development in the mouse: mosaics

In chapter 13 we described some of the ways in which mutant genes could be used in analyzing mammalian development in the usual situation in which every cell of the body has the same genotype. But we have seen in chapter 15 that in insects a very powerful technique is provided by the production—by somatic crossing over, somatic mutation or by loss through nondisjunction at mitosis of one of the sex-chromosomes in a line of cells—of embryos consisting of two genetically distinct cell populations;

B. Diagram of the embryo and extra-embryonic membranes at this stage, sectioned along the midline.
Partly based upon R. Rugh.

such individuals are called *mosaics*. In the mouse the only mosaics it is possible to make use of in practice are those which arise through loss of an X-chromosome, and in fact such mosaicism occurs normally in any female mammal which is heterozygous for any gene which is sex-linked. This is explained by the hypothesis proposed by Lyon in 1961 (and see her review of 1972) that, in order to achieve the same number of X-linked genes in females (XX) as in males (XY), one of the X-chromosomes in each cell of the embryo at an early stage (probably soon after implantation) becomes inactivated, and the same chromosome is inactive in all of its descendants; thereafter the organism will consists of a mixture of patches of cells of two different genotypes, which may be expressed phenotypically—as in the case of the coat of the tortoiseshell cat. An example in the mouse is provided by the gene *Tabby (Ta)*; females which are homozygous for this gene have a uniformly coloured velvety (because the long coarse guard hairs are not produced) coat, but in heterozygotes the coat is striped.

Mouse chimaeras

There is another way in which organisms of mixed genotype may be produced, developed independently by Tarkowski and Mintz in the 1960s and now established as an extremely important research technique. This is to combine cells from two embryos of different parentage to form a single embryo, and this can be achieved in the most spectacular way by removing the zona pellucida, usually at the 8-cell stage, and pushing the embryos together; once stuck, the cells rearrange themselves to form a single large morula which is then transferred to the uterus of another mouse. Such is the capacity for regulation that development proceeds completely normally, even the size being adjusted at the time of implantation. Given the number of mutant genes available in the laboratory mouse, it is easy to see what opportunities for developmental studies are opened up.

Genetic analysis still cannot be on just the same lines as in *Drosophila*, since cells in vertebrates move around and mingle and interact with each other to an extent which insect cells do not, creating a difficulty but at the same time an opportunity for investigating different types of problems. A more practical difficulty is the rarity of mutants which produce easily identifiable yet cell-autonomous phenotypic effects. The experiments described involve the melanocytes of the hairs where, though the pigment is secreted into the intercellular space, the effect is reasonably localized. Many more examples of the uses of the technique are provided in McLaren's review of 1976.

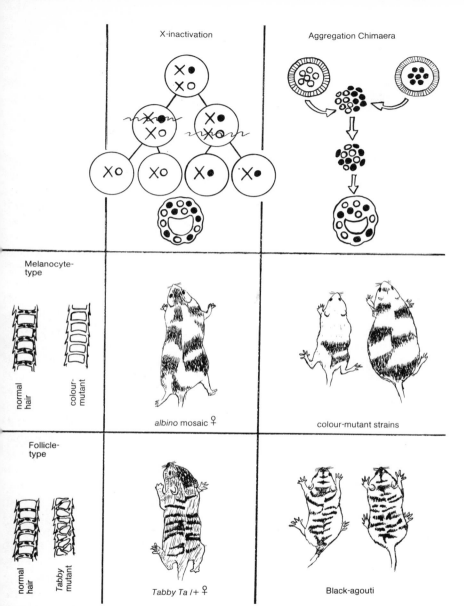

Figure 16.4 Coat-pigment patterns in mice derived from embryos whose cells are of mixed genotype, through genetic mosaicism arising from inactivation of the X-chromosome, or through chimaerism produced by combining cells from two embryos. Pigment differences result from differences in distribution of the pigment cells (melanocyte type) or from disturbances in the structure of the hairs (follicle type).
Based upon A. McLaren.

Coat pigment patterns in chimaeric and mosaic mice

The hairs arise as cylinders of keratin produced by the hair follicles, which are pits formed by impushings of ectodermal cells with at the base a mesodermal component—the dermal papilla. We have seen that the melanocytes are descendants of melanoblasts which migrate out from the neural crest and colonize the follicles, where they secrete pigment into the hair. Genes may control the type or amount of pigment produced, or the form in which it is laid down, either through their activity in the melanocytes directly, or indirectly through their activity in the epidermal follicle cells; the appearance in chimaeras of these melanocyte patterns and follicle patterns is quite different.

Take first an example of the melanocyte type. In 1967(a) Mintz produced chimaeras combining albino and various colour genes, and showed that in the resulting mice the colours appeared in broad transverse stripes on the head, body and tail, with a sharp discontinuity of pattern at the dorsal midline. Her explanation, summarized in 1971(b) is that each band represents a clone of melanoblasts descended from a single progenitor cell in the neural crest, established between 5 and 7 days of gestation before the neural folds have met and fused, and that the transverse striping reflects the route of migration of the melanoblasts between 8 and 12 days. There appear to be 34 primordial melanoblasts, a chain of 17 on each side, giving a maximum of 17 bands—3 on the head, 6 on the body and 8 on the tail but, since it is a matter of chance whether the same or different coloured primordial cells are adjacent to one another, the actual number of visible bands is usually less. Where hairs of different colour are mixed—and this is usually found at the edges of the bands—the coat is said to be *brindled*, and McLaren and Bowman showed in 1969 that chimaerism may extend to the level of individual hairs, in which the follicles have been colonized by two different melanoblast types.

The transverse bands appear to be regions within which melanoblasts move freely, but between which little mingling takes place in normal development. If this is the case, these regions would constitute something very like the compartments found in insect development (see chapter 15), suggesting that this concept may be widely applicable in development. In 1972 Lyon pointed out that if two or more primordial melanoblasts populated each region, single coloured bands would still arise through chance association of melanoblasts of the same pigment type; this would explain the frequent occurrence of bands which are brindled across the whole of their width, but the number could not be very large. We are left with the problem of how the regional boundaries are established, and how only one or a few primordial melanoblasts are allocated to each region.

Now consider an example of the follicle type. The *tabby* gene acts on pigmentation through its effect on the hair follicle by altering the structure of the hair, and chimaeras produced by combining homozygous *tabby* and normal embryos produce a striped pattern which is much finer than the banding patterns found in melanocyte chimaeras. Mintz calculated that follicle patterns might be produced as clones arising from about 85 precursor cells along the longitudinal axis on each side of the dorsal midline and suggested, since in regions where somites were clearly developed their number corresponded approximately with the number of bands, that a precursor cell arose in each somite, producing the dermal papillae of hairs in that region, and affecting their epidermal components indirectly.

A difficulty with this hypothesis is that, since the bands also appear on the head, it requires the additional postulate that transient somites which do not reach the level of morphological visibility are formed in the head mesoderm, and there is no evidence for this. McLaren suggests as an alternative explanation that the hair follicle bands do not represent localized clones of contrasting genotype but rather reflect some underlying threshold phenomenon superimposed on a wave pattern which arises from the presence and interaction of two genetically distinct cell populations.

It is possible that there is some underlying developmental mechanism linking the hair follicle pattern with the striped coat patterns which occur as normal species characteristics in many mammals and which have always presented a tantalizing challenge to developmental biologists. Turing (see chapter 12) is supposed to have said, regarding the development of the zebra, that the stripes were easy—only the horse part was difficult, but nobody has yet (though in 1977 Bard made an interesting analysis based on a generated wave pattern) provided a completely convincing model. In the zebra at least, a relation with the somites appears to be ruled out, since the stripes occur not only on the head but in transverse rings around the limbs.

CHAPTER SEVENTEEN

DEVELOPMENTAL NEUROBIOLOGY

THIS SUBJECT POSES SPECIAL PROBLEMS WHICH ARE AMONG THE MOST interesting and certainly the most challenging in developmental biology. They arise from the characteristic features of the nervous system, consisting as it does essentially of cells (neurones) which ramify throughout the body by fine cytoplasmic extensions (neurites), the longest of which (axons) may be up to a metre in length in mammals; whose functioning depends upon the precision with which connections between these cells are established in a system of literally indescribable complexity; and whose adjustment in its fine details probably continues as long as learning and memory-building persists. Though some invertebrates, especially insects, provide especially advantageous material for investigating some of its aspects, we shall follow it only in vertebrates. The earliest stages, leading to the formation of the neural tube and the neural crest have been described in chapter 7.

Derivatives of the neural tube

At the time of closure, the neural tube consists in most vertebrates of a single layer of neuroepithelial cells—the germinal cells, though it gives the appearance of consisting of two layers of cells of different type. This appearance arises because of a remarkable movement of the nuclei in the course of the cell cycle, so that metaphase occurs with the nuclei adjacent to the surface of the cell next to the tube lumen and the DNA synthesis stage when it is at the opposite end.

The germinal cells give rise to neuroblasts, more strictly immature neurones, since they are no longer capable of cell division, and precursors of the neuroglia or glial cells. The latter have a supporting function in the nervous system and those produced from ectoderm in this way are of two types, astrocytes and oligodendrocytes; a third type is of mesodermal origin and arises by transformation of macrophages which enter nervous tissue from the blood. It is likely, though not absolutely established, that neurones and glial cells arise from different germinal stem cells. Astrocytes and dendrocytes, on the other hand, probably arise from the same cell lines, and intermediate forms are found. Both are characterized by much-branched cytoplasmic processes extending in all directions from the cell body, but there are fewer of these in the oligodendrocytes.

The wall of the neural tube rapidly becomes multilayered as the various precursor cells lose their attachment to the membrane of its lumen and move into what is then known as the *mantle layer*. The innermost layer of proliferating germinal cells is now called the *ependymal layer*. To the outside a third layer, the *marginal layer*, becomes distinguished, occupied by the cytoplasmic extensions from the neuroblasts in the mantle layer which will become their axons. If it is going to make contact with some peripheral effector organ, rather than with another neurone in the central nervous system, i.e. if it is from a somitic motor neurone rather than an association neurone, the axon emerges from the neural tube and continues its outgrowth into the tissues. Outgrowth of dendrites, and establishment of their connections by synapses with axons of association neurones, does not occur until the axon has made contact with its peripheral connection, which is most commonly a muscle.

Derivatives of the neural crest

As we have seen in chapter 7, not all cells deriving from the neural crest contribute to the nervous system; cells arising from it are found in a wide variety of tissues and have in common only their site of origin and the fact that they have migrated long distances to their ultimate position. The neural derivatives of the crest are the neurones of the sensory (dorsal root) ganglion, neurones of the sympathetic ganglia and probably the Schwann cells. Schwann cells were directly observed by R. G. Harrison in 1904, migrating proximo-distally along nerve fibres in the developing tadpole tail. In 1924 he established (though Weston's more recent autoradiographic labelling studies do not exclude the possibility of a neural origin) that they originated only from the neural crest by removing the latter early in development and showing that, though the peripheral nerves continue to grow out normally, they do not become myelinated. In the central nervous system, the same function is performed by the oligodendrocytes. Schwann cells become attached to the axon shortly after it emerges from the neural tube or from the dorsal ganglion, and they continue to divide by mitosis, providing most of the sheath cells in this way. Proliferation appears to be controlled by interaction with the axon, since exactly the right number of

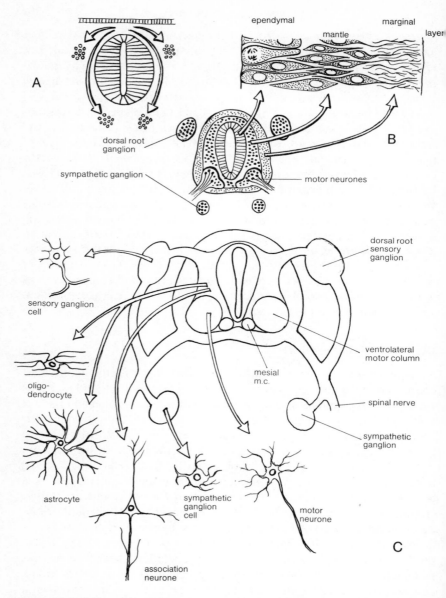

Figure 17.1 Differentiation of cells of the neural tube and the neural crest in vertebrates.
A. Early formation of the spinal cord from the neural tube, and migration of neural-crest cells to form the sensory ganglia of the dorsal roots and the sympathetic ganglia of the autonomic nervous system.

Schwann cells is produced to envelope each axon. Whether or not they go on to produce a myelin sheath is also determined by the axon, as Simpson and Young demonstrated in 1945 in regenerating myelinated and unmyelinated fibres. The process of myelination remained mysterious until Geren and Raskind, using the relatively new technique of electron microscopy in 1953, showed that the sheath is formed by an extension of the cell wrapping itself spirally around the axon, then squeezing out the cytoplasm so that only the protein-lipid cell membranes remain.

Morphogenesis of the neural tube

The neural tube is, from its mode of formation from a more or less pear-shaped neural plate, larger anteriorly than posteriorly; further regional differences appear as morphogenesis of the spinal cord and the complex structures of the brain proceeds. These changes of form occur chiefly through the organized activities of its component cells, notably through regional differences in the extent of germinal cell proliferation or of neuroblast death, and degeneration changes in size and shape, and in movements of whole neuroblasts or of their cytoplasmic extensions. We have some knowledge of the factors by which these activities are controlled, chiefly by cell-cell interactions, including interactions between the nerve cells and peripheral tissue cells, from which an understanding of the mechanisms underlying the integrated development of the nervous system begins to emerge. As an example of a comparatively simple and well-studied aspect we take the development of motor neurones in the spinal cord.

Relation of the development of motor neurones in the spinal cord to the muscles they innervate

In the chick embryo, the development of motor neurones has been described by Levi-Montalcini in 1950. They arise in the ependymal layer of the basal plate and, from day 2 and starting at the anterior end of the cord, they migrate to form a compact ventro-lateral motor column on each side, which at 4 days is of uniform thickness along the cord. At 5 days the column separates into a smaller mesial column, from which the axial muscles will be

---

B. Differentiation of the ependymal, mantle and marginal layers within the spinal cord.
C. Differentiation of cells of the neural tube and the neural crest, and the establishment of nerve concentrations and pathways in and around the spinal cord.
Partly based upon J. Z. Young.

innervated, and a lateral column for limb muscle innervation. The regions of the lateral column at the levels of the wings and legs become enlarged through growth of the cell bodies; elsewhere, the lateral column disappears, through death and degeneration of the cells in the neck region, and partly also through migration mesially (to form visceral motor neurones) in the thorax and lumbar regions. Throughout this time, cell proliferation by mitosis has been going on, and most of the definitive neurones are present by 8 days, but far more motor neurones are produced than are required.

Axons grow out from these neurones, and some have reached the developing limb muscles by 5 days. Dendrites appear slightly later; sensory and association neurones have also been undergoing corresponding development, so that reflex arcs are established and become active. If appropriate connections with muscle fibres that have not already been supplied are not made, neurones degenerate, and it is chiefly in this way that the relationship between demand and supply is established. This relationship has been analyzed most completely by Hughes in 1961 and Prestige in 1967 in *Xenopus*, but the essential discovery was made in the chick by Hamburger in 1934 when he showed that removing a wing bud at $2\frac{1}{2}$ days leads to the death of between 30 and 60% of motor neurones in this region 3 days later. In 1939 he further showed that implanting an extra limb bud at $2\frac{1}{2}$ days, providing it is within reach of outgrowing fibres, leads to a corresponding increase in the number of mature neurones. It appears that control of the numbers of motor neurones in the spinal cord is achieved by overproduction along its whole length, with subsequent adjustment by degeneration where they serve no purpose. The early events—proliferation, early differentiation and axon outgrowth—are autonomous, but whether a neurone continues to differentiate and function, or dies and degenerates, depends on its axon making proper connections with a peripheral muscle cell, and consequently on some interaction between them.

Growth of the nerve axon

Since one of the most vital features of neural development is the establishment of proper connections of neurones with effector organs, or sense organs, or other neurones, often over long distances by way of the nerve axons, the manner of axon growth and how it is directed is obviously of great importance. At the beginning of the century it was not even known whether the axon was formed by fusion of a chain of originally independent cells or by outgrowth from a single cell. The technique of tissue culture was initiated by R. G. Harrison between 1907 and 1910 to test these alternatives, and he established the truth of the outgrowth theory by observing it directly in neurones from the neural tube of amphibian embryos. Since then, research has been concerned with the factors controlling and directing outgrowth in axons and dendrites. The key structure is the growth cone at the tip of the outgrowth, which in tissue culture appears as an expansion at the tip, with many very fine radiating cytoplasmic processes or filopodia. Ciné films made by Bray in 1970 show that these processes are in continual motion, advancing, dividing, retracting, starting again until a contact with a cell body or a dendrite becomes stable. The growth cone moves forward as the axon or dendrite lengthens, and resembles and may serve the same function as the ruffled membrane of a moving fibroblast in culture. It is extremely closely attached to the substrate, and may bifurcate several times. As it moves forward, more axon is laid down between it and the cell body, and most probably molecules synthesized in the cell body are being assembled in the growth cone and incorporated in axonal membrane, much in the same way as Abercrombie and Harris have proposed that membrane assembly takes place at the ruffled membranes in fibroblasts, in which, however, it is supposed to be broken down posteriorly, so that the whole cell moves forward rather than just the tip.

Orientation of the growing axon

We do not know how specific connections between the growth cones and their targets are established—whether by chemotaxis directly, or by selection of particular contacts after connection by many branches to a great variety of terminations. Before this proximity is reached, however, axons must have been guided somehow at least to the general region of the target. One factor which must play at least some part is contact guidance by mechanical factors, demonstrated in 1934 by Paul Weiss in tissue culture where nerve axons growing in a stretched plasma clot follow lines of stress in the substrate. The suggestion was that axons in the embryo followed cell walls or collagen fibres or some other track to the target cell, but it is very doubtful whether the simplicity of the situation in culture is at all comparable to the tortuous 3-dimensional system presented in the embryonic situation.

It is more likely that the major part is played by chemical interactions between the growth cone and the cells it bypasses and the cell it ultimately connects with. These interactions may be very diverse and almost everything remains to be discovered about them. Whether the interactions are all short-range, involving so-called chemoaffinity, or may be long-

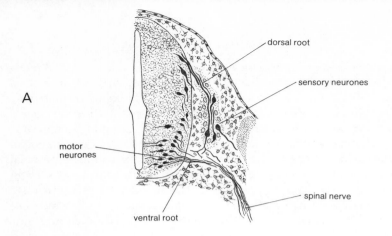

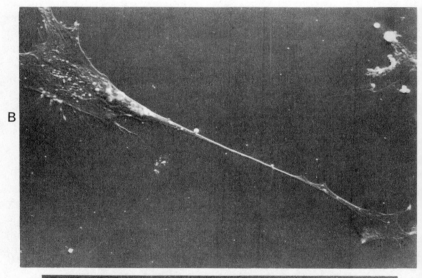

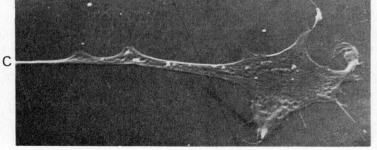

range, involving actual chemotaxis, is not even established. The clearest evidence for the latter was produced by DeLong and Coulombre in 1968, when they tested small pieces of 4-day chick retina on the surface of the tectum of 6- or 7-day embryos, giving an opportunity for the retino-tectal connection by way of the optic nerve fibres to be established between them. Whatever the position of the retinal transplant, the fibres from the graft grew preferentially to the tectal region it would have connected with in normal circumstances. But the results did not really eliminate the possibility of an initially random outgrowth, followed by selection of only those that happened to contact the appropriate tectal cells.

Morphogenesis of the brain

The most complex neural development occurs in the brain, formed from the anterior portion of the neural tube, where elaborate variations of the processes we have described above produce the large co-ordinating nerve centres such as the cerebral hemispheres and the cerebellum. Here we shall consider only the latter, the centre in the hind brain which controls the movements and balance of the body, receiving for this purpose afferent fibres from the inner ear and proprioceptor organs in the muscles. Impulses are relayed through a complex system of association neurones, directed to the motor centres of the midbrain, and then via the spinal cord to appropriate motor neurones innervating the muscles.

The cerebellar rudiment is formed (at 3–6 days of incubation in the chick) by migration of germinal cells from the ependymal layer of the neighbouring medulla over the surface of the cerebellum to form there a transient fourth layer—the external granular layer, which produces the granule cells and some glial cells. Neuroblasts produced in its own ependymal layer move out into the mantle layer and there differentiate to give the Purkinje cells, which form an outer layer which when fully developed is only one cell thick, and an inner layer of neurones which form

---

Figure 17.2
A. Transverse section of the neural tube of a pig embryo in which some of the neurones have been defined by a silver stain, showing how the dorsal root is formed from outgrowing axons from the neurones of the dorsal root ganglion, and the ventral root from axons which grow out from neurones of the lateral motor column.
Based upon J. Z. Young after Held.
B. Stereoscan electron-micrograph of a neurone in a tissue culture made from cells of the dorsal root ganglion of the mouse. *Top left*—cell body with nucleus; *bottom right*—growth cone.
C. The growth cone at higher magnification ($\times 1250$).

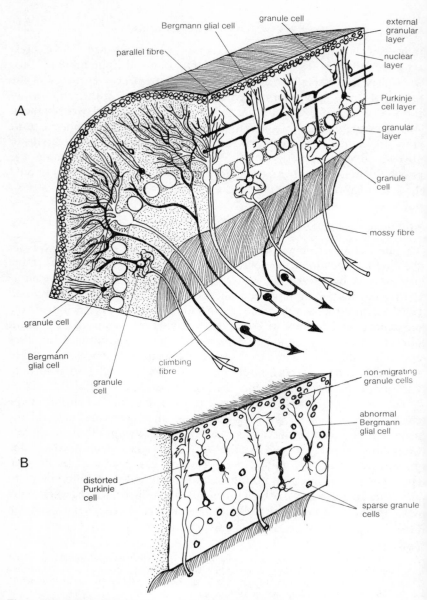

Figure 17.3 Stereo-diagram of part of the cerebellum of the mouse.
A. In a normal mouse. This highly complex neural network serves to amplify the input received from the incoming nerve fibres by relaying them through vast numbers of granule and Purkinje cells.
Based partly upon J. Z. Young.

the cerebellar roof nuclei. The Purkinje cell is of a very characteristic form; its dendrites grow outwards and resemble the branches of a tree trained in espalier fashion—fanwise and all in one plane, which is the same for all cells, occupying what becomes the "nuclear" layer. Its axon grows inwards, through what becomes the "granular" layer to a cerebellar nucleus where it synapses with an association neurone in the nucleus, whose axon goes to the midbrain or the medulla, but which has leading off from it a recurrent fibre (the climbing fibre) which leads back to the dendrites of the Purkinje cell where it becomes much divided to form many synaptic junctions with those dendrites. The granular layer becomes filled with granule cells which have migrated inwards from the outer granular layer and whose dendrites connect with the incoming fibres (mossy fibres) from the inner ear and the proprioreceptors; the cell's axon, which is payed out as the cell moves down into the granular layer, becomes T-shaped, the cross-piece becoming enormously elongated, forming one of the "parallel" fibres which run at right angles through the dendritic trees of the Purkinje cells and establish synaptic junctions with them as they pass at special projecting sites known as *spines*. The function of this complex neural net is apparently to amplify the input from the relatively few incoming mossy fibres, by relaying them through the granule cells and the Purkinje cells to produce so-called avalanche conduction of nerve impulses.

Genetic analysis of neuronal and glial cell development in the cerebellum

The morphogenetic analysis of anything as complex as the cerebellum is impossible by the techniques of conventional experimental embryology, but many behavioural mutants exist in mice which provide a means of dissecting the neurodevelopmental process, as in the investigations by Sidman, who reviewed them in 1974, and others. These studies all show a high degree of interaction between neurones in development, particularly in the dependence for the development of one type on its establishing normal synaptic connections with neurones of another type. Thus, in the *weaver* mouse mutant there is a gross reduction in the number of mature granule cells, though the neuroblasts are produced in normal numbers in the outer granular layer. In 1973 Rakic and Sidman suggested that the

---

. In a *weaver* mutant mouse. The granule cells—either because of an autonomous gene effect or because the Bergmann glial cell ladder is distorted—cannot establish themselves in the granular layer. This leads, through developmental interactions with the nerve-cell types, to wide-ranging disturbances of neuronal structure and function.
Based upon P. Rakic and R. L. Sidman.

primary action of the gene is not upon the granule cells but upon the Bergmann glial cells; they form ladder-shaped fibres perpendicular to the surface, which the granular-cell neuroblasts spiral down in their migration to the granular layer and which, though present in normal numbers, often look abnormal in the *weaver* mutant. On this hypothesis the granule cells would die as a result of not reaching their normal tissue environment, but in 1974 Sotelo and Changeux found that in *weaver* mice with a different genetic background even the few granule cells which do complete the migration die, suggesting that there is an autonomous lethal effect in this cell. No other cell types are affected directly, but the form of the dendritic tree of the Purkinje cells is very distorted, giving it a weeping-willow appearance, through disruption of the cells' layers caused by the failure of migration and death of the granule cells. But although the parallel fibres are not produced to establish a connection with them, the receptor surfaces on the Purkinje cell dendrites—the spines—are formed in normal numbers. In the absence of their normal synaptic association with the parallel fibres these Purkinje dendrite spines become occupied by the terminations of the mossy fibres whose normal targets are the granule cells, indicating that the details of the neural circuits are established through cellular interaction and that in unusual circumstances abnormal connections may be formed.

Retino-tectal connections and neuronal specificity

We have seen in chapter 9 that the retina is formed from the optic vesicle which arises from the wall of the brain, so that the optic nerve is in fact a cerebral tract rather than a true peripheral nerve. Thus the axon fibres of the retinal ganglion cells pass along the optic stalk directly to the visual centre in the midbrain—the optic tectum, crossing at the optic chiasma so that fibres from the eye on one side of the head will connect with neurones on the opposite side of the tectum. This constitutes a sensory system of very high resolution, in which the development of the neural circuits must be achieved in a precisely controlled way; it has been the subject of intensive investigation, especially since the work of Sperry, summarized in 1963 introduced the concept of neuronal specificity.

Sperry's original experiments were done on the adult toad, in which he showed that, if the optic nerve is severed or crushed, the nerve will regenerate and the toad will recover its ability to flick its tongue in the right direction to catch flies or other food objects in its visual field. This is remarkable enough, since there are many thousands of nerve fibres, each of which has to connect with a particular region of the tectum—even perhaps a particular neurone, and at the regenerating surface these fibres will be tangled in all directions; each has therefore to find its way to precisely the right place in order to make the proper functional connections. But more than that, if the eye is rotated through 180° before the nerve is severed, the connections appropriate to the original orientation of the retina are re-established, so that when function is restored by regeneration the toad will flick out its tongue in the direction which is precisely opposite to the location of its prey. This suggested that, contrary to the then-prevalent view that neuronal circuits are established by a process of adaptation through function in an initially random network, there exist precise chemical affinities between the neurones in the retinal system and those in the tectal system which enable fibres growing out from the one to locate and establish synaptic contact with specific cell bodies in the other; this is the hypothesis of neuronal specificity. Each cell carries a label, most likely cytochemical in nature, and matches up with a cell in the other system carrying a corresponding label.

More precision has been introduced into recent work through use of the technique of electrophysiological mapping, in which the tectum of the organism—usually the amphibian *Xenopus* or the goldfish—is exposed and a grid superimposed on it, on which electrodes may be applied at any point to record the arrival of nerve impulses. A spot of light is then moved about in the visual field on a device which enables its location to be identified on a corresponding grid system until a recording is obtained in the tectum. In this way corresponding points on the retina and the tectum—presumed to represent the relation between retinal ganglion cells and the synapses between their fibre terminations and neurones in the tectum—can be mapped. The terminology used may seem confusing at first, e.g. some authors refer to the visual field while others refer to the retinal field, and the relation of these to the tectal field is of course reversed (see figure 17.4); and direction in the antero-posterior axis is referred to in special terms. The tectum is a bilobed structure which lies transversely across the roof of the midbrain, and in a normal animal retinal ganglion cells connect with cells of the tectal lobe of the opposite side as follows: nasal (anterior) retinal cells to caudal (posterior) tectal cells; temporal (posterior) retinal cells to rostral (anterior) tectal cells; dorsal retinal cells to lateral tectal cells; ventral retinal cells to medial tectal cells. For the visual field: the antero-posterior axis corresponds with that of the tectum; the dorso-ventral axis corresponds with the mesio-lateral axis of the tectum. The mapping is continuous in both axes across the grids, suggesting that these specificities may be determined as a response to gradients of positional information in

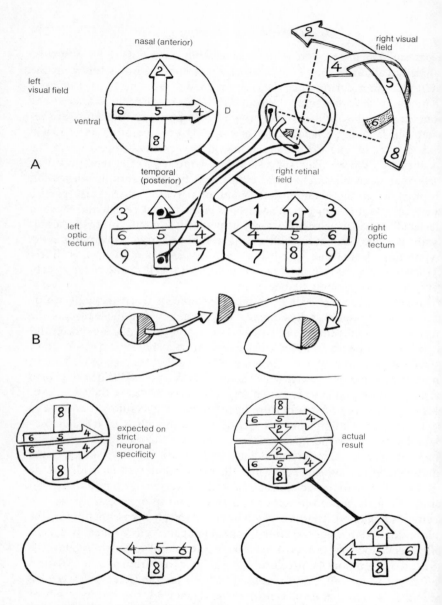

Figure 17.4 Establishment of retinal connections with the optic tectum in the visual system of *Xenopus*.
A. The normal system of retino-tectal connections. *Right*: the relation of the visual field to the retinal field of the right eye, with retinal ganglion cells connected by their neurites to the cells of the left optic tectum (optic lobe) of the midbrain, passing up the optic nerve and

development in the same way as other specific states of determination arise in other cell types. We have seen that there are states of determination which are not expressed by differentiation as a particular tissue type, but that decide whether, for example, a cell should interact with other cells to produce wing rather than leg; and consequently that cells must carry labels, determined by their position in the embryo, which produce these specific differences. How far this specificity extends, or is necessary for development, is not known—it is possible but extremely unlikely that every cell in the body carries a different label but, if the hypothesis of neuronal specificity is correct, it is clear that in the development of the nervous system the labelling resolution is very fine.

In 1968 Jacobson established (by rotating the eye through 180° in the embryo at various stages before retino-tectal connections had been established and noting some weeks later, when the connections were formed, whether the mapping to the tectum was as in the normal embryo or reversed and/or inverted) that specification occurs in the retinal ganglion cells at the tail-bud stage in *Xenopus* within a short period (5–7 hours) and in two stages; first the antero-posterior polarity is fixed and then the dorso-ventral axis, in a way which recalls the determination of polarity in these axes in the developing amphibian limb. It appeared as if the theory of neuronal specificity was essentially correct and that, once they had arrived at the tectum, the retinal fibres (probably by exploratory activity of the growth cone) sought out their proper partners among the tectal cells and, having found them, established synaptic connection with them; the specificities would be determined as a response to positional information in the developing retina and tectum, and the matching of cell specificities in the two systems would have been arrived at through natural selection in the course of evolution.

However, it soon became apparent that there were two difficulties in the way of this straightforward model. First there was the question of growth: in the *Xenopus* retinal rudiment at the time of Jacobson's operations there are only about 500 ganglion cells, but in the adult there are about 50 000,

---

crossing to the opposite side at the optic chiasma. *Left*: showing how the visual field may be mapped onto the tectum by electrophysiological mapping.
Based upon M. Jacobson.
B. Substitution of the anterior half of the retina from one side of a post-tailbud embryo for the posterior half of the retina from the other side, giving a "double-nasal" compound eye. The results which might be predicted if neuronal specificities were fixed are shown to the left, and the results actually obtained are shown to the right.
Based upon R. M. Gaze, M. Jacobson and G. Szèkely.

the accretion being by mitosis at the circular margin of the developing retina; if the pattern of specification is rigidly determined for each cell and inherited by its descendants, it is difficult to see how the pattern is retained in the retina as a whole. Moreover, the tectum grows in a different way, by accretion at one end, so that the only way for the two systems to remain matched throughout development would be for the synaptic relations to be constantly changing.

The second difficulty arises from the results of experiments in which compound eyes were formed, e.g. consisting of two nasal (anterior) half-retinas joined together, produced by removing the temporal (posterior) half of one optic rudiment and replacing it with the nasal half from the opposite side at a stage when determination of both axes had been completed. It might be expected, if their specificities are fixed, that cells of the posterior parts of the retina would only establish connections with their partners in the anterior part of the tectum, and that this would be indicated by electrophysiological mapping at maturity. But this is not what happens; instead, the fibres from each nasal half-retina establish connections across the whole of the antero-posterior extent of the tectum, retaining the order and polarity that would normally have been observed in the half-tectum. Thus a sort of regulation has occurred, each half-retina behaving as a whole retina and forming corresponding connections with the tectum (though the difficulties arising from growth patterns apply equally here), and this might be held to occur through respecification as a response to new positional information. But in this case the operation has been performed after determination has occurred, and in the goldfish a similar result was obtained when performed on the adult.

Neuronal interactions and systems matching

In 1972 Gaze and Keating questioned whether, when respecification can occur so readily and into adult life, the concept of specificity is useful or valid, and suggested instead the concept of systems-matching, implying that the retina and tectum match up not by exclusive cell-to-cell specificity but as systems. The difference at the level of actual cellular activity would be that, whereas in the neuronal specificity hypothesis each retinal fibre seeks out a uniquely labelled corresponding tectal neurone, in systems-matching the retinal fibres sort out among themselves into a pattern within the tectum, and that this pattern is orientated in relation to the tectum. What is determined in *Xenopus* at the tail-bud stage are the polarities of the retinal axes, which are always preserved thereafter and which provide the basis for establishing the correct pattern of connections of its fibres with the tectum later. Thus in 1976 Hope, Hammond and Gaze proposed a model which does not require the tectum to be specified, but in which any two retinal fibres which find themselves close to one another on the tectum must compare the positions of their cell bodies in the retina, but do not require to know their distance apart—only the direction of one from the other.

Whereas the neuronal specificity theory suggests there is very little interaction between cells of the same system, in the systems-matching theory both normal development and regulation following disturbance depend upon this. Other models for the development of neural networks which put a similar emphasis on neuronal interactions have been proposed, notably by Prestige and Willshaw in 1976, and by Willshaw and von der Malsburg, and Changeux and Danchin, in 1976; though highly speculative, they all point forward to an exciting era when developmental biology and neural biology, both concerned with the genesis of order and integration in the activity of cells—in neurology in producing ordered perceptions and responses, and ultimately in giving the sense of a unified personality; in development in producing coherently ordered complex processes and structures within embryos whose overall unity is always maintained—will be fully integrated fields of scientific endeavour and achievement.

FURTHER READING

General References
The following books treat various aspects of development in greater depth:
Balinsky, B.I. (1975). *An Introduction to Embryology*, 4th ed. Saunders: Philadelphia and London—especially good on organogenesis.
Bellairs, R. (1971). *Developmental Processes in Higher Vertebrates*. Logos Press, London.
Berrill, N. J. & Karp, G. (1976). *Development*. McGraw-Hill Book Company.
Billett, F. S. & Wild, A. E. (1975). *Practical Studies of Animal Development*. Chapman and Hall, London—extremely detailed advice on the maintenance of a wide variety of organisms and on experimental techniques for practical course-work in developmental biology.
Deuchar, E. M. (1975). *Cellular Interactions in Animal Development*. Chapman and Hall, London.
Fulton, C. & Klein, A. O., Eds. (1976). *Explorations in Developmental Biology*. Harvard University Press, Cambridge, Mass. and London—an excellent collection of many of the most significant research papers in the field, including both classical and contemporary work.
Graham, C. F. & Wareing, P. F., Eds. (1976). *The Developmental Biology of Plants and Animals*. Blackwell Scientific Publications, Oxford.
Lash, J. & Whittaker, J. R., Eds. (1974). *Concepts of Development*. Sinauer Assoc. Inc. Stamford, Conn.
Maclean, N. (1977). *The Differentiation of Cells*. Arnold, London.
Nelson, O. E. (1953). *Comparative Embryology of the Vertebrates*. McGraw-Hill Book Company, New York.
Poste, G. & Nicolson, G. L., Eds. (1976). *The Cell Surface in Animal Embryogenesis and Development*. North-Holland Publ. Co., Amsterdam, New York, Oxford.
Wessells, N. K. (1977). *Tissue Interactions and Development*. W. A. Benjamin, Inc., Menlo Park, Calif.

Chapter 1
Overview
Bonner, J. T. (1974). *On Development*. Harvard University Press, Cambridge.
Gasking, E. (1967). *Investigations into Generation 1651–1828*. Hutchinson, London.

Bonner, J. T. (1967). *The Cellular Slime Moulds*, 2nd ed., Princeton University Press.
Bonner, J. T. (1976). "Signalling systems in *Dictyostelium*." In: *The Developmental Biology of Plants and Animals* (Graham, C. F. and Wareing, P. F., Eds.) pp. 204–215. Blackwell, London.
Gerisch, G. (1968). "Cell aggregation and differentiation in *Dictyostelium*." In: *Current Topics in Developmental Biology*, Vol. 3 (Moscona, A. A. and Monroy, A., Eds.), pp. 157–197. Academic Press, London.

Hämmerling, J. (1963). "Nucleo-cytoplasmic interactions in *Acetabularia* and other cells." *Ann. Rev. Plant Physiol.* **14**, 65–92.
Harris, H. (1970) *Nucleus and Cytoplasm* 2nd ed., Clarendon Press, Oxford.
Jacob, F. & Monod, J. (1961). "On the regulation of gene activity," Cold Spring Harbor Symp., Quant. Biol. **26**, 193–211.
Kuroda, Y. (1970). "Differentiation of ommatidium-forming cells of *Drosophila melanogaster* in organ culture." *Expl. Cell Res.* **59**, 429–439.
Morgan, T. H. (1934). *Embryology and Genetics*, Columbia University Press, New York.
Shelton, P. M. J. (1976). "The development of the insect compound eye." In: *Insect Development* (Lawrence, P.A., Ed.) pp. 152–169, Blackwell, Oxford.
Waddington, C. H. (1940). *Organisers and Genes*, Cambridge University Press.
Waddington, C. H. & Perry, M. M. (1960). "The ultra-structure of the developing eye of *Drosophila*." *Proc. Roy. Soc. B.* **153**, 155–178.
Watson, D. & Crick, F. H. (1953). "Genetical implications of the structure of deoxyribonucleic acid." *Nature*, Lond., **171**, 964–967.
Wilson, E. B. (1925). *The Cell in Development and Heredity*, 3rd ed. Macmillan, New York.
Wilson, E. O. (1971). *The Insect Societies*, Harvard University Press, Cambridge, Massachusetts.

Chapter 2
Overview
Austin, C. R. and Short, R. V., Eds. (1972). *Reproduction in Mammals, Book 1. Germ Cells and Fertilization.* Cambridge University Press.
Longo, F. J. & Anderson, E. (1974). "Gametogenesis." In: *Concepts of Development* (Lash, J. & Whittaker, J. R., Eds.), Sinauer, Stamford, Conn.
Raven, C. P. (1961). *Oogenesis: The Storage of Developmental Information.* Pergamon Press.

Barros, C. & Austin, C. R. (1967). "*In vitro* fertilization and the sperm acrosome reaction in the hamster." *J. exp. Zool.* **166**, 317–323.
Brown, D. D. & Dawid, I. B. (1968). "Specific gene amplification in oocytes." *Science* **160**, 272–280.
Brown, E. H. & King, R. C. (1964). "Studies on the events resulting in the formation of an egg chamber in *Drosophila melanogaster.*" *Growth* **28**, 41–81.
Colwin, L. H. & Colwin, A. L. (1964). "Role of the gamete membranes in fertilization." Soc. Study Devl. Growth Symp. 22, Academic Press.
Counce, S. J. (1963). "Developmental morphology of polar granules in *Drosophila*." *J. Morph.* **112**, 129–145.
Davidson, E. H. (1977). *Gene Activity in Early Development*, 2nd ed. Academic Press, New York.
Davidson, E. H. & Hough, B. R. (1971). "Genetic information in oocyte RNA." *J. Mol. Biol.* **56**, 491–506.
Dubois, R. (1967). "Localisation et migration des cellules germinales du blastoderme non incubé de Poulet, d'après les résultats de cultures *in vitro.*" *Archs. Anat. micr. Morph. exp.* **56**, 245–264.
Dubois, R. (1968). "La colonisation des ébauches gonadiques par les cellules germinales de l'embryon de Poulet, en culture *in vitro.*" *J. Embryol. exp. Morph.* **20**, 189–213.
Fawcett, D. W. (1970). "A comparative view of sperm ultrastructure. *Biol. Reprod. suppl.* **2**, 90–127.
Gill, K. S. (1963). "Developmental genetic studies on oogenesis in *Drosophila melanogaster*." *J. exp. Zool.* **152**, 251–278.
Kille, R. A. (1960). "Fertilization of the lamprey egg." *Exp. Cell Res.* **20**, 12–27.
King, R. C., Rubinson, A. C. & Smith, R. F. (1956). "Oogenesis in adult *Drosophila melanogaster.*" *Growth* **20**, 121–157.

Krause, G. (1961). "Preformed ooplasmic reaction systems in insect eggs." Symp. on Germ Cells and Development, pp. 302–337. Fond. Baselli, Ravia.
MacGregor, H. C. (1972). "The nucleolus and its genes in amphibian oogenesis." *Biol. Rev.* **47**, 177–210.
Miller, R. L. (1966). "Chemotaxis during fertilization in the Hydroid, *Campanularia*." *J. exp. Zool.* **162**, 23–44.
Mintz, B. & Russell, E. S. (1957). "Gene-induced embryological modifications of primordial germ cells in the mouse." *J. exp. Zool.* **134**, 207–237.
Olsen, M. W. (1960). "Nine year summary of parthogenesis in turkeys." *Proc. Soc. Exp. Biol.* **105**, 279–281.
Roosen-Runge, E. C. (1962). "The process of spermatogenesis in mammals." *Biol. Rev.* **37**, 343–377.
Smith, L. D. (1966). "The role of a germinal plasm in the formation of primordial germ cells in *Rana pipiens*." *Devl. Biol.* **14**, 330–347.
Smith, L. D. & Williams, M. A. (1975). "A critique and review of germ plasm and primordial germ-cell determination in insects and vertebrates." In: *Biology of Reproduction* (Markert, C. L. & Papaconstantinou, J., Eds.), Academic Press.
Ullmann, S. L. (1973). "Oogenesis in *Tenebrio molitor*: Histological and autoradiographical observations on pupal and adult ovaries." *J. Embryol. exp. Morph.* **30**, 179–217.
Whitten, W. K. (1971). "Parthogenesis: does it occur spontaneously in mice?" *Science* **171**, 406–407.

Chapter 3
Overview
Davidson, E. H. (1977). *Gene Activity in Early Development*, 2nd ed. Academic Press, New York.
Gurdon, J. B. (1974). *The Control of Gene Expression in Animal Development*. Clarendon Press, Oxford.

Balinsky, B. L. (1966). "Changes in ultrastructure of amphibian eggs following fertilization." *Acta Embryol. Morph. Exper.* **9**, 132–154.
Briggs, R. & King, T. J. (1952). "Transplantation of living nuclei from blastula cells into enucleated frog's eggs." *Proc. Nat. Acad. Sci. USA* **38**, 455.
Clement, A. C. (1962). "Development of *Ilyanassa* following removal of the D macromere at successive cleavages." *J. exp. Zool.* **149**, 193–215.
Conklin, E. G. (1905). "Organisation and cell-lineage in the Ascidian egg." *J. Acad. Nat. Sci. Phila.* **13**, 2.
Curtis, A. S. G. (1960). "Cortical grafting in *Xenopus laevis*." *J. Embryol. exp. Morph.* **8**, 163–173.
Graham, C. F. (1966). "Regulation of DNA synthesis and mitosis in multinucleate frog eggs." *J. Cell Sci.* **1**, 363–372.
Graham, C. F., Arms, K. & Gurdon, J. B. (1966). "The induction of DNA synthesis by frog egg cytoplasm." *Devl. Biol.* **14**, 349–382.
Graham, C. F. & Morgan, R. W. (1966). "Changes in the nature of the cell cycle during early amphibian development." *Devl. Biol.* **14**, 439–460.
Gurdon, J. B. (1962). "The developmental capacity of nuclei taken from intestinal epithelium cells of feeding tadpoles." *J. Embryol. exp. Morph.* **10**, 622–640.
Gurdon, J. B. & Woodland, H. R. (1968). "The cytological control of nuclear activity in animal development." *Biol. Rev.* **43**, 233–267.
Harris, H. (1970). *Cell Fusion*. Clarendon Press, Oxford.
Jaffé, L. F. (1968). "Localization in the developing *Fucus* egg and the general role of localizing currents." *Advan. Morphog.* **7**, 295–328.
Raven, C. P. and Bezem, J. J. (1973). "Computer simulation of embryonic development. IV.

Normal development of the *Lymnaea* egg." *Proc. Koninkl. Nederl. Akad. van Wetenschappen-Amsterdam C,* **76**, 23–35.

Rünnstrom, J. (1966). "The vitelline membrane and cortical particles in sea urchin eggs and their function in maturation and fertilization." *Advan. Morphog.* **5**, 221–325.

Tucker, J. B. & Meats, M. (1976). "Microtubules and control of insect egg shape." *J. Cell Biol.* **71**, 207–217.

Wilson, E. B. (1904). "Experimental studies on germinal localization. I. The germ-regions in the egg of *Dentalium.*" *J. exp. Zool.* **1**, 1–72.

Zalokar, M. & Erk, I. (1976). "Division and migration of nuclei during early embryogenesis of *Drosophila melanogaster.*" *J. Microscop. Biol. Cellulaire* **25**, 97–106.

Chapter 4
Overview

Davidson, E. H. (1977). *Gene Activity in Early Development,* 2nd ed., Academic Press, New York.

Gurdon, J. B. (1974). *The Control of Gene Expression in Animal Development,* Clarendon Press, Oxford.

Counce, S. J. (1956). "Studies on female-sterility genes in *Drosophila melanogaster.* I. The effects of the *gene deep* orange on embryonic development." *Z. induct. Abst. Vererb. Lehr.* **87**, 443–461.

Driesch, H. (1891). "Entwicklungsmechanische studien. I. Der werth der beiden ersten furchungszellen in der echinodermenentwicklung." *Z. Wiss. Zool.* **53**, 160–178. Translated as: "The potency of the first two cleavage cells in echinoderm development." In: *Foundations in Experimental Embryology* (Willier, B. H. and Oppenheimer, J. M., Eds.) 1974, Prentice-Hall.

Garen, A. & Gehring, W. (1972). "Repair of the lethal developmental defect in *deep orange* embryos of *Drosophila* by injecting egg cytoplasm." *Proc. natn. Acad. Sci. USA* **69**, 2982–2985.

Hörstadius, S. (1973). *Experimental Embryology of Echinoderms,* Clarendon Press, Oxford.

Kalthoff, K. (1975). "Specification of the antero-posterior body pattern in insect eggs." In: *Insect Development* (Lawrence, P. A., Ed.), pp. 53–75. Blackwell Scientific, Oxford.

Poulson, D. F. (1945). "Chromosomal control of embryogenesis in *Drosophila.*" *Amer. Nat.* **79**, 340–363.

Sander, K. (1975). "Pattern specification in the insect embryo." In: *Cell Patterning,* CIBA Symp. **29** (new series), pp. 241–263.

Wright, T. R. F. (1970). "The genetics of embryogenesis in *Drosophila.*" *Advan. Genet.* **15**, 261–395.

Zalokar, M., Audit, C. & Erk, I. (1975). "Developmental defects of female sterile mutants of *Drosophila melanogaster.*" *Devl. Biol.* **47**, 419–432.

Chapter 5
Overview

CIBA Symp. 14 (new series) (1973). *Locomotion of Tissue Cells.* North-Holland, Amsterdam.

Trinkaus, J. P. (1969). *Cells into Organs.* Prentice-Hall.

Abercrombie, M. & Heaysman, J. E. M. (1953). "Observations on the social behaviour of cells in tissue culture. I. Speed of movement of chick heart fibroblasts in relation to their mutual contacts." *Exptl. Cell Res.* **5**, 111–131.

Abercrombie, M. & Ambrose, E. J. (1958). "Interference microscope studies of cell contacts in tissue culture." *Exptl. Cell Res.* **15**, 332–345.

Abercrombie, M. & Ambrose, E. J. (1962). "The surface properties of cancer cells. A review." *Cancer Res.* **22**, 525.

Armstrong, P. B. & Armstrong, M. T. (1973). "Are cells in solid tissues immobile? Mesonephric mesenchyme studied *in vitro.*" *Devl. Biol.* **35**, 187–209.
Bard, J. B. L. & Hay, E. D. (1975). "The behaviour of fibroblasts from the developing avian cornea. Morphology and movement *in situ* and *in vitro.*" *J. Cell Biol.* **67**, 400–418.
Curtis, A. S. G. & van de Vyver, G. (1971). "The control of cell adhesion in a morphogenetic system." *J. Embryol. exp. Morph.* **26**, 295–312.
Garrod, D. R. (1973). "Tissue-specific sorting-out in two dimensions in relation to contact inhibition of cell movement." *Nature* **244**, 568–569.
Holtfreter, J. (1947). "Observations on the migration, aggregation and phagocytosis of embryonic cells." *J. Morphol.* **80**, 25–55.
Huxley, J. S. (1911). "Some phenomena of regeneration in *Sycon.*" *Phil. Trans. Roy. Soc. Lond.* B **202**, 165.
Ingram, V. M. (1969). "A side view of moving fibroblasts." *Nature* **243**, 445–449.
Ludueña, M. A. & Wessells, N. K. (1973). "Cell locomotion, nerve elongation and microfilaments." *Devl. Biol.* **30**, 427–440.
Minchin, E. A. (1908). "Materials for a monograph of the *Ascons.*" *Quart. J. Micr. Sci.* **52**, 301–355.
Moscona, A. A. (1952). "Cell suspensions from organ rudiments of chick embryo." *Expl. Cell Res.* **3**, 536–539.
Rosenberg, M. D. (1963). "Cell guidance by alterations in monomolecular films." *Science* **139**, 411–412.
Steinberg, M. S. (1973). "Cell movement in confluent monolayers: A re-evaluation of the causes of 'contact inhibition'." *Locomotion of Tissue Cells*, Ciba Foundation Symp. 14 (new series), North-Holland, Amsterdam.
Weiss, P. (1961). "Guiding principles in cell locomotion and cell aggregation." *Exptl. Cell Res. Suppl.* **8**, 269–281.
Weiss, P. & Taylor, A. C. (1960). "Reconstitution of complete organs from single-cell suspensions of chick embryos in advanced stages of differentiation." *Proc. natn. Acad. Sci.* **46**, 1177–1185.
Wilson, H. V. (1907). "On some phenomena of coalescence and regeneration in sponges." *J. exp. Zool.* **5**, 245–258.
Wourms, J. P. (1972). "The developmental biology of annual fishes. II. Naturally occurring dispersion and reaggregation of blastomeres during the development of annual fish eggs." *J. exp. Zool.* **182**, 169–200.

Chapter 6
Overview
Trinkaus, J. P. (1969). *Cells into Organs*. Prentice-Hall Inc. New Jersey.

Bellairs, R. (1971). *Developmental Processes in Higher Vertebrates*. University of Miami Press, Coral Gables, Florida.
Downie, J. R. & Pegrum, S. M. (1971). "Organisation of the chick blastoderm edge." *J. Embryol. exp. Morph.* **26**, 623–635.
Dubois, R. (1967). "Localisation et migration des cellules germinales du blastoderme non incubé de Poulet, d'après les résultats de cultures *in vitro*. *Archs. Anat. micr. Morph. exp.* **56**, 245–264.
Ede, D. A. & Counce, S. J. (1956). A cinematographic study of the embryology of *Drosophila melanogaster. Wilhelm Roux Arch. Entw mech. Org.* **148**, 402–415.
Gustafson, T. & Wolpert, L. (1967). "Cellular movement and contact in sea urchin morphogenesis." *Biol. Rev.* **44**, 442–498.
Holtfreter, J. (1943). "A study of the mechanics of gastrulation; Part I." *J. exp. Zool.* **94**, 261–318.

Holtfreter, J. (1949). "A study of the mechanics of gastrulation: Part II. *J. exp. Zool.* **95**, 171–212.
Keller, R. E. & Schoenwolf, G. (1976). "A scanning electron microscopic study of cellular morphology, contact and arrangement in the gastrula of *Xenopus laevis*." Reported by Trinkaus, J. P.: "On the mechanism of metazoan cell movements." In: *The Cell Surface in Animal Embryogenesis and Development* (Poste, G. & Nicolson, G. L., Eds.), pp. 225–329. North-Holland Publishing Company.
Lütz, H. (1949). "Sur la production expérimentale de la polyembryonie et de la monstruosité double chez les Oiseaux." *Arch. Anat. Micr. Morph. Exper.* **38**, 79–144.
Nakatsuji, N. (1975). "Studies on the gastrulation of amphibian embryos: light and electron microscopic observation of a urodele *Cynops pyrrhogaster*." *J. Embryol. exp. Morph.* **34**, 669–685.
Okazaki, K. (1960). "Skeleton formation of sea urchin larvae. II. Organic matrix of the spicule." *Embryologia* **5**, 283–320.
Olsen, M. W. (1962). "Polyembryony in unfertilized turkey eggs." *J. Hered.* **53**, 125–129.
Olsen, M. W. (1965). "Delayed development and atypical cellular organization in blastodiscs of unfertilized turkey eggs." *Devl. Biol.* **12**, 1–14.
Patterson, J. F. (1909). "Gastrulation in the pigeon's egg." *J. Morph.* **20**, 65.
Rosenquitz, G. C. (1971). "The origin and movements of the hepatogenic cells in the chick embryos as determined by radioautographic mapping." *J. Embryol. exp. Morph.* **25**, 97–113.
Sander, K. (1976). "Morphogenetic movements in insect embryogenesis." In: *Insect Development* (Lawrence, P. A., Ed), pp. 35–52 R. E. S. Symp. 8. Blackwell Scientific Publications, Oxford.
Smith, L. D. & Ecker, R. E. (1970). "Uterine suppression of biochemical and morphogenetic events in *Rana pipiens*." *Devl. Biol.* **22**, 622–637.
Spratt, N. T. Jr. & Haas, H. (1960*a*). "Morphogenetic movements in the lower surface of the unincubated and early chick blastoderm." *J. exp. Zool.* **144**, 139–158.
Spratt, N. T. Jr. & Haas, H. (1960*b*). "Importance of morphogenetic movements in the lower surface of the young chick blastoderm." *J. exp. Zool.* **144**, 257–276.
Tilney, L. G. & Gibbons, J. R. (1969). "Microtubules in the formation and development of the primary mesenchyme in *Arbacia punctulata*. II. An experimental analysis of their role in development and maintenance of cell shape." *J. Cell Biol.* **41**, 227–250.
Trinkaus, J. P. (1976). "On the mechanism of metazoan cell movements." In: *The Cell Surface in Animal Embryogenesis and Development* (Poste, G. & Nicholson, G. L., Eds.), Elsevier/North-Holland Biom. Press, Amsterdam.
Vakaet, L. (1962). "Some new data concerning the formation of the definitive endoblast in the chick embryo." *J. Embryol. exp. Morph.* **10**, 38–57.
Vollmar, H. (1972*a*). "Die einrollbewegung (Anatrepsis) des keimstreifs im ei von *Acheta domesticus* (Orthopteroidea, Gryllidae). *Wilhelm Roux Arch. Entw. mech. Org.* **170**, 135–151.
Vollmar, H. (1972*b*). "Frühembryonale gestaltungsbewegungen im vitalgefärbten dotter-entoplasma-system intakter und fragmentierter eier von *Acheta domesticus* L. (Orthopteroidea)." *Wilhelm Roux Arch. Entw. mech. Org.* **171**, 288–243.

Chapter 7
Baker, P. C. & Schroeder, T. E. (1967). "Cytoplasmic filaments and morphogenetic movements in the amphibian neural tube." *Devl. Biol.* **15**, 432–450.
Bancroft, M. & Bellairs, R. (1975). "Differentiation of the neural plate and neural tube in the young chick embryo. A study by scanning and transmission electron microscopy." *Anat. Embryol.* **147**, 309–335.
Bellairs, R. (1963). "The development of somites in the chick embryo." *J. Embryol. exp. Morph.* **11**, 697–714.

FURTHER READING

Bellairs, R., Breathnach, A. S. & Gross, M. (1975). "Freeze-fracture replication of junctional complexes in unincubated and incubated chick embryos." *Cell Tiss. Res.* **162**, 235–252.
Burnside, B. (1971). "Microtubules and microfilaments in newt neurulation." *Devl. Biol.* **26**, 416–441.
Byers, B. & Porter, K. R. (1964). "Oriented microtubules in elongating cells of developing lens rudiment after induction." *Proc. Nat. Aca. Sci.* USA **52**, 1091–1098.
Cloney, R. A. (1966). "Cytoplasmic filaments and cell movements: epidermal cells during ascidian metamorphosis." *J. Ultrastruct. Res.* **14**, 300–328.
Farquhar, M. G. & Palade, G. E. (1963). "Functional complexes in various epithelia." *J. Cell Biol.* **17**, 375–412.
Flint, O. P. (1977). "Cell interactions in the developing axial skeleton in normal and mutant mouse embryos." In: *Vertebrate Limb and Somite Morphogenesis* (Ede, D. A., Hinchliffe, J. R. and Balls, M., Eds.), pp. 465–484. British Society for Developmental Biology Symposium 3, Cambridge University Press, Cambridge.
Furshpan, E. J. & Potter, D. D. (1968). "Low resistance junctions between cell in embryos and tissue cultures." *Curr. Top. Devl. Biol.* **3**, 95–127.
Hay, E. D. (1973). "Origin and role of collagen in the embryo." *Amer. Zool.* **13**, 1085–1109.
Hay, E. D. & Meier, S. (1974). "Glycosaminoglycan synthesis by embryonic inductors: neural tube, notochord and lens." *J. Cell Biol.* **62**, 889–898.
Hörstadius, S. (1950). *The Neural Crest*, Oxford University Press, New York & London.
Jacobson, A. G. & Gordon, R. (1976). "Changes in the shape of the developing vertebrate nervous system analyzed experimentally, mathematically and by computer simulation." *J. Exp. Zool.* **197**, 191–246.
Karfunkel, P. (1974). "The mechanisms of neural tube formation." *Int. Rev. Cytol.* **38**, 245–271.
Langman, J. & Nelson, G. R. (1968). "A radioautographic study of the development of the somite in the chick embryo." *J. Embryol. exp. Morph.* **19**, 217–226.
Le Douarin, N. (1976). "Cell migration in early vertebrate development studied in interspecific chimeras." In: *Embryogenesis in Mammals*, Ciba Foundation Symp. 40 (new series). pp. 71–101. Elsevier/Excerpta/North-Holland, Amsterdam.
Lipton, B. H. & Jacobson, A. G. (1974). "Analysis of normal somite development." *Devl. Biol.* **38**, 73–90.
Lowenstein, W. R. (1973). "Membrane junctions in growth and differentiation." *Fed. Proc., Fed. Amer. Soc. Exp. Biol.* **32**, 60–64.
Mayer, T. C. (1973). "The migratory pathway of neural crest cells into the skin of mouse embryos." *Devl. Biol.* **34**, 39–46.
O'Hare, M. J. (1973). "A histochemical study of sulphated glycosaminoglycans associated with the somites of the chick embryo." *J. Embryol. exp. Morph.* **29**, 197–208.
Pearce, T. L. & Zwann, J. (1970). "A light and electron microscopic study of cell behaviour and microtubules in the embryonic chicken lens using Colcemid." *J. Embryol. exp. Morph.* **23**, 491–507.
Revel, J-P., Chang, L. L. & Yip, P. (1973). "Cell junctions in the early chick embryo — a freeze etch study." *Devl. Biol.* **35**, 302–317.
Sanders, E. J. (1973). "Intercellular contact in the unincubated chick embryo." *Z. Zellforsch* **141**, 459–468.
Sheridan, J. D. (1976). "Cell coupling and cell communication during embryogenesis." In: *The Cell Surface in Animal Embryogenesis and Development* (Poste, G. & Nicolson, G. L., Eds.), Elsevier/North-Holland Biomedical Press, Amsterdam.
Spratt, N. T., Jr. (1955). "Analysis of the organizer center in the early chick embryo. I. Localization of prospective notochord and somite cells." *J. exp. Zool.* **128**, 121–164.
Subak-Sharpe, H., Burke, R. R. & Pitts, J. D. (1966). "Metabolic cooperation by cell to cell transfer between genetically marked mammalian cells in tissue culture." *Heredity* **21**, 342–343.

Teillet, M. A. & Le Douarin, N. (1970). "La migration des cellules pigmentaires étudiée par la méthode des greffes heterospecifiques de tube nerveux chez l'embryon d'Oiseau." *C. R. Acad. Sci. Ser. D* **270**, 3095–3098.
Trelstad, R. L., Hay, E. D. & Revel, J-P. (1967). "Cell contact during early morphogenesis in the chick embryo." *Devl. Biol.* **16**, 78–106.
Twitty, V. C. (1966). *Of Salamanders and Scientists*, Freeman, San Francisco.
Twitty, V. C. & Niu, M. C. (1954). "The motivation of cell migration studied by the isolation of embryonic pigment cells singly and in small groups *in vitro*." *J. exp. Zool.* **125**, 541–574.
Wessells, N. K. *et al.* (1971). "Microfilaments in cellular and developmental processes." *Science* **171**, 135–143.
Weston, J. A. (1963). "A radioautographic analysis of the migration and localization of trunk neural crest cells in the chick." *Devl. Biol.* **6**, 279–310.

Chapter 8
Overview
Maclean, N. (1977). *The Differentiation of Cells*, Arnold, London.

Allison, A. C. (1959). "Recent developments in the study of inherited anaemias." *Eugen. Q.* **6**, 155–164.
Cahn, R. D. & Cahn, M. B. (1966). "Heritability of cellular differentiation in retinal pigment cells *in vitro*." *Proc. Nat. Acad. Sci. USA* **55**, 104–114.
Cole, R. J. & Tarbutt, R. G. (1973)." Kinetics of cell multiplication and differentiation during adult and prenatal haemopoiesis." In: *The Cell Cycle in Development and Differentiation* (Balls, M, and Billet, F. S., Eds.), Cambridge University Press, Cambridge.
Eguchi, G. (1976). "'Transdifferentiation' of vertebrate cells in cell culture." In: *Embryogenesis in Mammals*, pp. 241–257, Ciba Foundation Symp. 40 (new series), Elsevier/Excerpta Medica/North-Holland, Amsterdam.
Eguchi, G., Abe, S-I, & Watanabe, K. (1974). "Differentiation of lens-like structures from newt iris epithelial cells *in vitro*." *Proc. Nat. Acad. Sci.* **71**, 5052–5056.
Farusawa, M., Ikawa, Y. & Sugano, H. (1971). "Development of erythrocyte membrane-specific antigens in clonal cultured cells of Friend virus-induced tumour." *Proc. Jap. Acad.* **47**, 220–224.
Fell, H. B. & Mellanby, E. (1953). "Metaplasia produced in cultures of chick ectoderm by high vitamin A." *J. Physiol.* **119**, 470–488.
Friend, C. (1957). "Cell-free transmission in adult swiss mice of a disease having the character of a leukaemia." *J. Exp. Med.* **105**, 307–318.
Harrison, P. R., Conkie, D. & Paul, J. (1973). "Role of cell division and nucleic acid synthesis in erythropoietin-induced maturation of foetal liver cells *in vitro*." In: *The Cell Cycle in Development and Differentiation*. (Balls, M., and Billett, F. S., Eds.), Cambridge University Press, Cambridge.
Holtzer, H., Weintraub, H., Mayer, R. & Mochran, B. (1972). "The cell cycle, cell lineages and cell differentiation." *Curr. Topics in Develop. Biol.* **7**, 229–256.
Konigsberg, I. R. (1963). "Clonal analysis of myogenesis." *Science* **140**, 1273–1284.
McCulloch, E. A. & Till, J. E. (1962). "The sensitivity of cells from normal mouse bone marrow to gamma radiation *in vitro* and *in vivo*." *Radiation Res.* **16**, 822–832.
Marks, P. A. & Kovach, J. S. (1966). "Development of mammalian erythroid cells." In: *Current Topics in Developmental Biology*, Vol. 1 (A. A. Moscona & A. Monroy, Eds.), pp. 213–252. Academic Press, New York.
Okada, T. S. *et al.* (1975) "Differentiation of lens in cultures of neural retinal cells of chick embryos." *Devl. Biol.* **45**, 318–329.
Paul, J. & Hickey, I. (1974). "Haemoglobin synthesis in inducible, uninducible and hybrid Friend cell clones." *Exp. Cell Res.* **87**, 20–30.

Waddington, C. H. (1940). *Organizers and Genes*, Cambridge University Press, Cambridge.
Weiss, P. (1939). *Principles of Development*, Holt, New York.

Chapter 9

Overview

Saxén, L., Karkinen-Jääskeläinen, M., Lehtonen, E., Nordling, S. and Wartiovaara, J. (1976). "Inductive tissue interactions." In: *The Cell Surface in Animal Embryogenesis and Development* (G. Poste and G. L. Nicolson, Eds.), Elsevier/North-Holland Biomedical Press, Amsterdam.

Wessells, N. K. (1977). *Tissue Interactions and Development*. W. A. Benjamin Inc., Menlo Park, California.

Barth, L. G. & Barth, L. J. (1972). "Sodium and calcium uptake during embryonic induction in *Rana pipiens*." *Devl. Biol.* **28**, 18–34.

Bernfield, M. R., Banerjee, S. D. & Cohen, R. H. (1972). "Dependence of salivary epithelial morphology and branching morphogenesis upon acid mucopolysaccharide-protein (proteoglycan) at the epithelial surface." *J. Cell Biol.* **52**, 674–689.

Dodson, J. W. (1967a). "The differentiation of epidermis. I. The interrelationship of epidermis and dermis in embryonic chicken skin." *J. Embryol. exp. Morph.* **17**, 83–105.

Dodson, J. W. (1967b). "The differentiation of epidermis. II. Alternative pathways of differentiation of embryonic chicken epidermis in organ culture." *J. Embryol. exp. Morph.* **17**, 107–117.

Ellison, M. L. & Lash, J. W. (1971). "Environmental enhancement of *in vitro* chondrogenesis." *Devl. Biol.* **26**, 486–496.

Fleischmajer, R. & Billingham, R. E., Eds. (1968). *Epithelial-Mesenchymal Interactions*. Williams and Wilkins. Baltimore.

Gossens, C. L. & Unsworth, B. R. (1972). "Evidence for a two-step mechanism operating during *in vitro* mouse kidney tubulogenesis." *J. Embryol. exp. Morph.* **28**, 615–631.

Hauschka, S. D. & Konigsberg, I. R. (1966). "The influence of collagen on the development of muscle clones." *Proc. Nat. Acad. Sci.* **55**, 119–126.

Hendrix, R. W. & Zwann, J. (1975). "The matrix of the optic vesicle-presumptive lens interface during induction of the lens in the chicken embryo." *J. Embryol. exp. Morph.* **33**, 1023–1049.

Holtfreter, J. (1945). "Neurulization and epidermization of gastrula ectoderm." *J. Exp. Zool.* **98**, 161–209.

Jacobson, A. G. (1958). "The roles of neural and non-neural tissues in lens induction." *J. Exp. Zool.* **139**, 528–558.

Kratochwill, K. (1969). "Organ specificity in mesenchymal induction demonstrated in the development of the mammary gland of the mouse." *Devl. Biol.* **20**, 46–71.

McKeehan, M. S. (1958). "Induction of portions of the chick lens without contact with the optic cup." *Anat. Rec.* **132**, 297–305.

McMahon, D. (1974). "Chemical messengers in development: a hypothesis." *Science* **185**, 1012–1021.

Needham, J. (1950). *Biochemistry and Morphogenesis*, Harvard University Press, Cambridge, Massachusetts.

Raven, C. P. (1958). "Abnormal development of the foregut in *Limnaea stagnalis*." *J. Exp. Zool.* **139**, 189–246.

Rutter, W. J., Pictet, R. L. & Morris, P. W. (1973). "Towards molecular mechanisms of developmental processes." *Ann. Rev. Biochem.* **42**, 601–646.

Saxén, L., Koskimies, O., Lahti, A., Miettinen, H., Rapola, J. and Wartiovaara, J. (1968). "Differentiation of kidney mesenchyme in an experimental model system." *Advances in Morphogenesis* **7**, 251–292.

Saxén, L. & Toivonen, S. (1961). "The two-gradient hypothesis in primary induction. The

combined effect of two types of inductors mixed in different ratios." *J. Embryol. exp. Morph.* **9**, 514–533.
Saxén, L. Toivonen, S. & Vainio, T. (1964). "Initial stimulus and subsequent interactions in embryonic induction." *J. Embryol. exp. Morph.* **12**, 333–338.
Spemann, H. (1901a). "Entwicklungsphysiologische Studien am Triton-Ei. I." *Wilhelm Roux Arch. Entw. mech. Org.* **12**, 224–264.
Spemann, H. (1901b). "Über Correlationen in der Entwicklung des Auges." *Anat. Anzeiger* **19**, Ergänzungsheft, 61–79.
Spemann, H. & Mangold, H. (1924). "Über induktion von embryonalanlagen durch implantation artfremder organisatoren." *Wilhelm Roux. Arch. Entw. mech. Org.* **100**, 599–638.
Tiedemann, H. (1968). "Factors determining embryonic differentiation." *J. Cell Physiol.* **72**, 129–144 (Suppl. 1).

Chapter 10
Overview
Sengel, P. (1976). *Morphogenesis of Skin*. Cambridge University Press.

Billingham, R. E. & Silvers, W. K. (1967). "Studies on the conservation of epidermal specificities of skin and certain mucosas in adult mammals." *J. Exp. Med.* **125**, 429–446.
Bullough, W. S. (1962). "The control of mitotic activity in adult mammalian tissues." *Biol. Rev.* **37**, 307–342.
Bullough, W. S. & Deol, J. U. R. (1971). "The pattern of tumour growth." In: *Control Mechanisms of Growth and Differentiation*. Symp. Soc. Expr. Biol. 25, Cambridge University Press.
Christophers, E. (1972). "Correlation between column formation, thickness and rate of new cell production in guinea pig epidermis." *Virchows Arch. Abt. B. Zellpath.* **10**, 286–292.
Claxton, J. H. (1964). "The determination of patterns with special reference to that of the central primary skin follicles in sheep." *J. Theoret. Biol.* **7**, 302–317.
Cohen, S. & Elliott, G. A. (1963). "The stimulation of epidermal keratinization by a protein isolated from the submaxillary gland of the mouse." *J. invest. Derm.* **40**, 1–6.
Cohen, S. & 'Espinasse, P. G. (1961). "On the normal and abnormal development of the feather." *J. Embryol. exp. Morph.* **9**, 223–251.
Coulombre, J. L. & Coulombre, A. J. (1971). "Metaplastic induction of scales and feather in the corneal anterior epithelium of the chick embryo." *Devl. Biol.* **25**, 464–478.
Ede, D. A. (1972). "Cell behaviour and embryonic development." *Int. J. Neuroscience* **3**, 165–174.
Ede, D. A., Hinchliffe, J. R. & Mees, H. C. (1971). "Feather morphogenesis and feather pattern in normal and talpid^3 chick embryos." *J. Embryol. exp. Morph.* **25**, 65–83.
Fell, H. B. & Mellanby, E. M. (1953). "Metaplasia produced in cultures of chick ectoderm of high Vitamin A." *J. Physiol. Lond.* **119**, 470–488.
Garber, B. B. & Moscona, A. A. (1964). "Aggregation *in vivo* of dissociated cells. I. Reconstruction of skin in the chorioallantoic membrane from suspensions of embryonic chick and mouse skin cells." *J. exp. Zool.* **155**, 179–202.
Gross, J. (1956). "The behaviour of collagen units as a model in morphogenesis." *J. Biophys. Biochem. Cytol. suppl.* **2**.
Harris, W. F. (1972). "Dislocations and control of integumentary patterns in the chick." *Dev. Biol.* **29**, f-1–f-4.
Iversen, O. H., Bjerknes, R. & Devik, F. (1968). "Kinetics of cell renewal, cell migration and cell loss in the hairless mouse dorsal epidermis." *Cell Tissue Kinet.* **1**, 351–367.
Linsenmayer, T. F. (1972). "Control of integumentary patterns in the chick." *Devl. Biol.* **27**, 244–271.

McKenzie, I. C. (1969). "Ordered structure of the stratum corneum of mammalian skin." *Nature.* **222**, 881–882.
McKenzie, I. C. (1970). "Relationship between mitosis and the ordered structure of the stratum corneum in mouse epidermis." *Nature.* **226**, 653–655.
Moscona, M. H. & Moscona, A. A. (1965). "Control of differentiation in aggregates of embryonic skin cells: suppression of feather morphogenesis by cells from other tissues." *Devl. Biol.* **3**, 402–423.
Novel, G. (1973). "Feather pattern stability and reorganization in cultured skin." *J. Embryol. exp. Morph.* **30**, 605–633.
Rawles, M. E. (1965). "Tissue interactions in the morphogenesis of the feather." In: *Biology of the Skin and Hair Growth* (Lyne, A. G. & Short, B. F., Eds.). Angus & Robertson, Sydney.
Sengel, P. (1971). "The organogenesis and arrangement of cutaneous appendages in birds." *Adv. Morphog.* **9**, 181–230.
Sengel, P. (1975). "Feather pattern development." In: *Cell Patterning* Ciba Foundation Symp. 29, pp. 51–70. Associated Scientific Publishers, Amsterdam.
Stuart, E. S. & Moscona, A. A. (1967). "Embryonic morphogenesis: Role of fibrous lattice in the development of feathers and feather patterns." *Science* **157**, 947–948.
Weiss, P. & Kavanau, J. (1957). "A model of growth and control in mathematical terms." *J. Gen. Physiol.* **41**, 1.
Wessells, N. K. (1965). "Morphology and proliferation during early feather development." *Devl. Biol.* **12**, 131–153.

Chapter 11
Overview
Ede, D. A., Hinchliffe, J. R. & Balls, M. M., Eds. (1977). *Vertebrate Limb and Somite Morphogenesis*, Cambridge University Press.

Abbot, U. & Holtzer, H. (1966). "The loss of phenotypic traits by differentiated cells. III. The reversible behaviour of chondrocytes in primary cultures." *J. Cell Biol.* **28**, 473–487.
Amprino, R. (1965). "Aspects of limb morphogenesis in the chicken." In: *Organogenesis* (DeHaan, R. & Ursprung, H., Eds.), pp. 225–281. Holt, Rinehart & Winston, New York.
Christ, B., Jacob, H. J. & Jacob, M. (1977). "Experimental analysis of the origin of the wing musculature in avian embryos." *Anat. Embryol.* **150**, 171–186.
Dienstman, S. R., Biehl, J., Holtzer, S. & Holtzer, H. (1974). "Myogenic and chondrogenic lineages in developing limb buds grown *in vitro*." *Dev. Biol.* **39**, 83–95.
Ede, D. A. and Law, J. T. (1969). "Computer simulation of vertebrate limb morphogenesis." *Nature,* **221**, 244–248.
Errick, J. E. & Saunders, J. W. Jr. (1974). "Effects of an "inside-out" limb-bud ectoderm on development of the avian limb." *Dev. Biol.* **41**, 338–351.
Globus, M, Vethamany-Globus, S. (1976). "An *in vitro* analogue of early chick limb bud outgrowth." *Differentiation* **6**, 91–96.
Harrison, R. G. (1918). "Experiments on the development of the forelimb of *Amblystoma*, a self-differentiating, equipotential system." *J. exp. Zool.* **25**, 413–462.
Hornbruch, A. & Wolpert, L. (1970). "Cell division in the early growth and morphogenesis of the chick limb." *Nature* **226**, 764–766.
Kieny, M. (1964a). "Etude du mécanisme de la régulation dans le developpement du bourgeon de membre de l'embryon de poulet. I. Regulation des excédents." *Dev. Biol.* **9**, 197–229.
Kieny, M. (1964b). "Etude du mécanisme de la régulation dans le developpement du bourgeon de membre de l'embryon de poulet. II. Regulation des déficiances dans les chimères 'aile-patte' et 'patte-aile'." *J. Embryol. exp. Morph.* **12**, 357–371.
Rubin, L. & Saunders, J. W. Jr. (1972). "Ectodermal-mesodermal interactions in the growth

of limb buds in the chick embryo: constancy and temporal limits of the ectoderma induction." *Devl. Biol.* **28**, 94–112.
Saunders, J. W. Jr. (1948). "The proximo-distal sequence of origin of the parts of the chic wing and the role of the ectoderm." *J. exp. Zool.* **108**, 363–403.
Saunders, J. W. Jr. & Fallon, J. F. (1966). "Cell death in morphogenesis." In: *Majo Problems in Developmental Biology* (Locke, M., Ed.) pp. 289–314. Academic Press, Ne York.
Saunders, J. W. Jr., Gasseling, M. T. & Saunders, L. C. (1962). "Cellular death i morphogenesis of the avian wing." *Devl. Biol.* **5**, 147–178.
Saunders, J. W. Jr. & Gasseling, M. T. (1968). "Ectodermal-mesenchymal interactions in th origin of limb symmetry." In: *Epithelial-Mesenchymal Interactions* (Fleischmajer, R. & Billingham, R. E., Eds.) pp. 78–97, Williams & Wilkins, Baltimore.
Slack, J. M. W. (1976). "Determination of polarity in the amphibian limb." *Nature* **26** 44–46.
Stark, R. J. & Searls, R. L. (1973). "A description of chick wing bud development and a mode of limb morphogenesis." *Devl. Biol.* **33**, 138–158.
Stocum, D. L. (1975). "Outgrowth and pattern formation during limb ontogeny an regeneration." *Differentiation* **3**, 167–182.
Summerbell, D., Lewis, J. H. & Wolpert, L. (1973). "Positional information in chick lim morphogenesis." *Nature* **244**, 492–496.
Toole, B. P. (1972). "Hyaluronate turnover during chondrogenesis in the developing chic limb and axial skeleton." *Devl. Biol.* **29**, 321–329.

Chapter 12
Overview
Davies, D. D. & Balls, M. M., Eds. (1971). *Control Mechanisms of Growth and Differentiatior* 25th Symposium of the Society for Experimental Biology. Cambridge University Press

Arbib, M. A. (1972). "Automata theory in the context of theoretical embryology." In *Foundations of Mathematical Biology*. Vol. II, Academic Press, New York & London.
Bard, J. & Lauder, I. (1974). "How well does Türing's theory of morphogenesis work?" *J. theor. Biol.* **45**, 501–531.
Child, C. M. (1928). "The physiological gradients." *Protoplasma* **5**, 447–476.
Cooke, J. & Zeeman, E. C. (1976). "A clock and wavefront model for control of the number c repeated structures during animal morphogenesis." *J. theor. Biol.* **58**, 455–476.
Fallon, J. F. & Crosby, G. M. (1977). "Polarizing zone activity in limb buds of amniotes." In *Vertebrate Limb and Somite Morphogenesis* (Ede, D. A., Hinchliffe, J. R. & Balls, M Eds.) pp. 55–69. British Society for Developmental Biology Symposium 3, Cambridg University Press, Cambridge.
French, V., Bryant, P. J. & Bryant, S. V. (1976). "A model for pattern regulation i epimorphic fields." *Science* **193**, 969–981.
Kauffman, S. A. (1973). "Control circuits for determination and transdetermination. *Science* **181**, 310–318.
McMahon, D. (1973). "A cell contact model for position determination in development. *Proc. Nat. Acad. Sci.* USA **70**, 2396–2400.
McMahon, D. & West, C. (1976). "Transduction of positional information durin development." In: *The Cell Surface in Animal Embryogenesis and Development* (Poste, G & Nicolson, G. L., Eds.), pp. 449–493. North Holland Publ. Co., Amsterdam, Ne York, Oxford.
Maynard, J. W. Jr., Smith, J. & Sonhi, K. C. (1961). "The arrangement of bristles i *Drosophila*." *J. Embryol. exp. Morph.* **9**, 661–672.
Saunders, J. W. Jr (1977). "The experimental analysis of chick limb bud development." In *Vertebrate Limb and Somite Morphogenesis* (Ede, D. A., Hinchliffe, J. R. & Balls, M

Eds.), pp. 1–24. British Society for Developmental Biology Symposium 3, Cambridge University Press, Cambridge.
Saunders, J. W. Jr. & Gasseling, M. T. (1968). "Ectodermal-mesenchymal interactions in the origin of limb symmetry." In: *Epithelial-Mesenchymal Interaction* (Fleischmajer, R. & Billingham, R. E., Eds.) Wiliams & Wilkins Co., Baltimore.
Tickle, C., Summerbell, D. & Wolpert, L. (1975). "Positional signalling and specification of digits in chick limb morphogenesis." *Nature* **254**, 199–202.
Turing, A. M. (1952). "The chemical basis of morphogenesis." *Phil. Trans. Roy. Soc. Lond.* **237**, 37–72.
von Neuman, J. (1951). "The general and logical theory of automata." In: *Cerebral Mechanisms in Behaviour: The Hixon Symposium*, pp. 1–32. Wiley, New York.
Wilby, O. K. & Ede, D. A. (1975). "A model generating the pattern of cartilage skeletal elements in the embryonic chick limb." *J. theor. Biol.* **50**, 199–217.
Wolpert, L. (1969). "Positional information and the spatial pattern of cellular differentiation." *J. theor. Biol.* **25**, 1–48.
Wolpert, L., Clark, M. R. B. & Hornbruch, A. (1972). "Positional signalling along *Hydra*." *Nature* **239**, 101–103.
Wolpert, L. & Lewis, J. H. (1975). "Towards a theory of development." *Federation Proc.* **34**, 14–20. F.A.S.E.B. Conf., Atlantic City, New Jersey, 1974.

Chapter 13
Overview
Suzuki, D. (1974). "Developmental genetics." In: *Concepts of Development* (Lash, J. & Whittaker, J. R., Eds.) pp. 349–379. Sinauer Assoc. Inc. Stamford, Conn.

Artz, K., Bennett, D. & Jacob, F. (1974). "Primitive teratocarcinoma cells express a differentiation antigen specified by a gene at the T-locus in the mouse." *P.N.A.S. USA* **71**, 811–814.
Bennett, D. (1964). "Abnormalities associated with a chromosome region in the mouse." *Science* **144**, 260–267.
Bennett, D. (1975). "The T-locus of the mouse." *Cell* **6**, 441–454.
Britten, R. J. & Davidson, E. H. (1969). "Gene regulation in higher cells: A theory." *Science* **165**, 349–357.
Ede, D. A., Bellairs, R. & Bancroft, M. (1974). "A scanning electron microscope study of the early limb bud in normal and *talpid*^{3} mutant chick embryos." *J. Embryol. exp. Morph.* **31**, 761–785.
Ede, D. A. & Flint, O. P. (1975a). "Intercellular adhesion and formation of aggregates in normal and *talpid*^{3} mutant chick mesenchyme." *J. Cell Sci.* **18**, 97–111.
Ede, D. A. & Flint, O. P. (1975b). "Cell movement and adhesion in the developing chick wing bud: studies on cultured mesenchyme cells from normal and *talpid*^{3} mutant embryos." *J. Cell Sci.* **18**, 301–314.
Ede, D. A. & Law, J. T. (1969). "Computer simulation of vertebrate limb morphogenesis." *Nature* **221**, 244–248.
Goldschneider, I. & Barton, R. W. (1976). "Development and differentiation of lymphocytes." In: *The Cell Surface in Animal Embryogenesis and Development* (Poste, G. and Nicolson, G. L., Eds.). pp. 599–695. North-Holland Publ. Co., Amsterdam, New York, Oxford.
Green, M. (1970). "Oncogenic viruses." *A. Rev. Biochem.* **39**, 701–756.
Grüneberg, H. (1963). *The Pathology of Development*, Blackwell, Oxford.
Jacob, F. & Monod, J. (1961). "Genetic regulatory mechanisms in the synthesis of protein." *J. Mol. Biol.* **3**, 318–356.
Mayer, T. C. (1973). "Site of gene action in *Steel* mice: analysis of the pigment defect by mesoderm-ectoderm recombinations." *J. exp. Zool.* **184**, 345–352.

Mayer, T. C. & Green, M. C. (1968). "An experimental analysis of the pigment defect caused by mutations at the *W* and *Sl* loci in mice." *Devl. Biol.* **18**, 62–75.
Pantelouris, E. M. (1968). "Absence of the thymus in a mouse mutant." *Nature* **217**, 370–371.
Poulson, D. F. (1945). "Chromosal control of embryogenesis in *Drosophila*." *Amer. Nat.* **79**, 340.
Stevens, L. C. (1967). "The biology of teratomas." *Adv. Morphogen.* **6**, 1–31.
Thompson, D'Arcy (1917). *On Growth and Form*. Cambridge University Press, Cambridge.
Waddington, C. H. (1957). *The Strategy of the Genes*. Allen & Unwin, London.

Chapter 14
Overview
Deuchar, E. M. (1975). "Long distance cellular interactions mediated by hormones." Chapter 10 in *Cellular Interactions in Animal Development*. Chapman & Hall, London.

Ashburner, M. & Richards, G. (1976). "The role of ecdysone in control of gene activity in polytene chromosomes of *Drosophila*." In: *Insect Development* (Lawrence, P. A., Ed.), Blackwell, Oxford, pp. 203–225.
Becker, H. J. (1962). "Die puffs der speicheldrüsenchromosomen von *Drosophila melanogaster*. II. Die auslösung der puffbildung, ihre spezifität und ihre beziehung zur funktion der ringdrüse." *Chromosoma* **13**, 341–384.
Chang, C. Y. & Witschi, E. (1956). "Genic control and hormonal reversal of sex differentiation in *Xenopus*." *Proc. Soc. exp. Biol. Med.* **93**, 140–144.
Clever, U. (1963). "Von der ecdysonkonzentration abhängige genaktivitätsmuster in den speicheldrüsen-chromosomen von *Chironomus tentans*." *Devl. Biol.* **6**, 73–98.
Doane, W. W. (1973). "Role of hormones in development." In: *Insects: Developmental Systems* (Counce, S. J. & Waddington, C. H., Eds.) Vol. 2, pp. 291–497. Academic Press, London.
Kratochwil, K. (1971). "*In vitro* analysis of the hormonal basis for the sexual dimorphism in the embryonic development of the mouse mammary gland." *J. Embryol. exp. Morph.* **25**, 141–153.
Kratochwil, K. (1976). "The androgen inhibitory effect on the mammary rudiment *in vitro*: development of responsiveness and tissue interaction in the response." In: "Mécanismes de la rudimentation des organes chez les embryons de vertébrés." *Colloques Int. CNRS* **266**, pp. 85–91.
Lillie, F. R. (1917). "The free-martin; a study of the action of sex hormones in the foetal life of cattle." *J. exp. Zool.* **23**, 371–452.
McMahon, D. (1974). "Chemical messengers in development: a hypothesis." *Science* **185**, 1012–1021.
O'Malley, B. W. *et al.* (1972). "Steroid hormone induction of a specific translatable messenger RNA." *Nature* New Biol. **240**, 45–47.
Schneiderman, H. A. & Gilbert, L. I. (1964). "Control of growth and development in insects." *Science* **143**, 325–333.
Short, R. V. (1970). "The bovine freemartin: a new look at an old problem." *Phil. Trans. Roy. Soc. Lond. B* **259**, 141–147.
Sláma, K. (1975). "Some old concepts and new findings on hormonal control of insect metamorphosis." *J. Insect Physiol.* **21**, 921–955.
Sutherland, E. W. & Rall, T. W. (1957). "The properties of an adenine ribonucleotide produced with cellular particles, ATP, Mg^{++}, and epinephrine or glucagon." *J. Am. Chem. Soc.* **79**, 3608.
Tata, J. A. (1971). "Hormonal regulation of metamorphosis." *Brit. Soc. exp. Biol. Symp.* **25**, on "Control Mechanisms of Growth and Differentiation."
Wigglesworth, V. B. (1961). "The epidermal cell." In *The Cell and the Organism* (Ramsay, J. A. & Wigglesworth, V. B., Eds.), Cambridge University Press, pp. 127–143.

Wigglesworth, V. B. (1976). "Juvenile hormone and pattern formation." In: *Insect Development* (Lawrence, P. A., Ed.), Symp. Roy. Ent. Soc. Lond. No. 8, pp. 186–202. Blackwell Scientific Publications, London.
Witschi, E. (1938). "Studies on sex differentiation and sex determination in amphibians. V. Range of the cortex-medulla antagonism in parabiotic twins of *Ranidae* and *Hylidae*." *J. exp. Zool.* **78**, 113–145.

Chapter 15
Overview
Counce, S. J. & Waddington, C. H., Eds. (1973). *Developmental Systems: Insects*, Vols. I and II. Academic Press, New York.
Lawrence, P. A., Ed. (1976). *Insect Development*, Blackwell, Oxford.

Bryant, P. J. (1974). "Determination and pattern formation in the imaginal discs of *Drosophila*. *Current Topics in Developmental Biology* (Moscona, A. A. & Monroy, A., Eds.) **8**, 41–80. Academic Press.
Bryant, P. J. & Schneiderman, H. (1969). "Cell lineage, growth and determination in the imaginal leg disc of *Drosophila melanogaster*." *Devl. Biol.* **20**, 263–290.
Crick, F. H. C. & Lawrence, P. A. (1975). "Compartments and polyclones in insect development." *Science* **189**, 340–347.
French, V., Bryant, P. J. & Bryant, S. V. (1976). "A model for pattern regulation in epimorphic fields." *Science* **193**, 969–981.
Garcia-Bellido, A., Ripoli, P. & Morata, G. (1973). "Developmental compartmentalisation of the wing disc of *Drosophila*." *Nature* New Biol. **245**, 251–253.
Gehring, W. (1967). "Clonal analysis of determination dynamics in cultures of imaginal discs in *Drosophila melanogaster*." *Devl. Biol.* **16**, 438–456.
Hadron, E. (1966). "Dynamics of determination." In: *Major Problems in Developmental Biology* (Locke, R. M., Ed.), pp. 85–104. Academic Press, New York.
Hotta, Y. & Benzer, S. (1972). "Mapping of behaviour in *Drosophila* mosaics." *Nature* **240**, 527–535.
Huxley, J. S. (1932). *Problems of Relative Growth*, Methuen, London.
Kauffman, S. A. (1973). "Control circuits for determination and transdetermination." *Science* **181**, 310–318.
Lawrence, P. A. (1966). "Gradients in the insect segment: the orientation of hairs in the milkweed bug *Oncopeltus fasciatus*." *J. exp. Biol.* **44**, 607–620.
Lawrence, P. A. (1973). "A clonal analysis of segment development in *Oncopeltus* (Hemiptera)." *J. Embryol. exp. Morph.* **30**, 681–699.
Lawrence, P. A. (1974). "Cell movement during pattern regulation in *Oncopeltus*." *Nature* **248**, 609–610.
Lawrence, P. A. & Morata, G. (1976). "The compartment hypothesis." In: *Insect Development* (Lawrence, P. A., Ed.), Roy. Ent. Soc. Symp. No. 8. Blackwell Scientific Publications, Oxford. pp. 132–149.
Lawrence, P. A., Crick, F. H. C. & Munro, M. (1972). "A gradient of positional information in an insect, *Rhodnius*." *J. Cell Sci.* **11**, 815–853.
Locke, M. (1959). "The cuticular pattern in an insect, *Rhodnius prolixus*." *J. exp. Biol.* **26**, 459–477.
Nardi, J. B. & Kafatos, F. C. (1976). "Polarity and gradients in lepidopteran wing epidermis. II. The differential adhesiveness model: gradient of a non-diffusible cell surface parameter." *J. Embryol. exp. Morph.* **36**, 489–512.
Poodry, C. A. & Schneiderman, H. A. (1970). "The ultrastructure of the developing leg of *Drosophila melanogaster*." *Wilhelm Roux. Arch. Entw. Mech. Org.* **166**, 1–44.
Russell, M. (1974). "Pattern formation in imaginal discs of a temperature-sensitive cell lethal mutant of *Drosophila melanogaster*." *Devl. Biol.* **40**, 24–39.

Shields, G. & Sang, J. H. (1970). "Characteristics of five cell types appearing during *in vitro* culture of embryonic material from *Drosophila melanogaster*." *J. Embryol. exp. Morph.* **23**, 53–69.
Waddington, C. H. (1940). *Organizers and Genes*, Cambridge University Press, Cambridge.
Whittle, J. R. S. (1976). "Mutations affecting the development of the wing." In: *Insect Development* (Lawrence, P. A., Ed.), pp. 118–131. Blackwell, Oxford.
Wieschaus, E. & Gehring, W. (1976). "Clonal analysis of primordial disc cells in the early embryo of *Drosophila melanogaster*." *Devl. Biol.* **50**, 249–263.
Wilson, E. O. (1971). *The Insect Societies*, Harvard University Press, Cambridge, Massachusetts.

Chapter 16
Overview
Sherman, M. I., Ed. (1977). *Concepts in Mammalian Embryogenesis*, MIT Press, Cambridge, Mass and London.
McLaren, A. (1976). *Mammalian Chimaeras*. Cambridge University Press.

Bard, J. B. L. (1977). "A unity underlying the different zebra striping patterns." *J. Zool. Lond.* **183**, 527–539.
Bellairs, R. (1971). *Developmental Processes in Higher Vertebrates*. University of Miami Press, pp. 366, Coral Gables, Florida.
Cole, R. J. (1967). "Cinemicrographic observations on the trophoblast and zona pellucida of the mouse blastocyst." *J. Embryol. exp. Morph.* **17**, 481–490.
Dalcq, A. M. (1957). *Introduction to General Embryology*, Oxford University Press.
Hillman, N., Sherman, M. I. & Graham, C. (1972). "The effect of spatial arrangement on cell determination during mouse development." *J. Embryol. exp. Morph.* **28**, 263–278.
Lyon, M. F. (1972). "X-chromosome inactivation and developmental patterns in mammals." *Biol. Rev.* **47**, 1–35.
Lyon, M. J. (1961). "Gene action in the X-chromosome of the mouse *(Mus musculus L.)*." *Nature* **190**, 372–373.
McLaren, A. & Bowman, P. (1969). "Mouse chimaeras derived from fusion of embryos differing by nine genetic factors." *Nature* **224**, 238–240.
Mintz, B. (1967). "Gene control of mammalian pigmentary differentiation. I. Clonal origin of melanocytes." *Proc. Nat. Acad. Sci. Wash.* **58**, 344–351.
Mintz, B. (1971). "Clonal basis of mammalian differentiation." In: *Control Mechanisms of Growth and Differentiation*, (Davies, D. D. and Balls, M., Eds.), pp. 345–369. Cambridge University Press.
Rugh, R. (1967). *The Mouse, its Reproduction and Development*. Burgess, Minneapolis.
Snell, G. D. & Stevens, L. C. (1966). "Early embryology." In: *Biology of the Laboratory Mouse*, 2nd ed. (Green, E. L., Ed.), pp. 205–245. McGraw-Hill, New York.

Chapter 17
Overview
Jacobson, M. (1970). *Developmental Neurobiology*. Holt, Rinehart and Winston.

Bray, D. (1970). "Surface movements during the growth of single explanted neurons." *Proc. Nat. Acad. Sci. USA* **65**, 905–910.
Changeux, J. P. & Danchin, A. (1976). "Selective stabilisation of developing synapses as a mechanism for the specification of neuronal networks." *Nature* **264**, 705–712.
DeLong, R. G. & Coulombre, A. J. (1968). "The specifity of retinotectal connections studied by retinal grafts onto the optic tectum in chick embryos." *Devl. Biol.* **16**, 513–531.

FURTHER READING

Gaze, M. & Keating, M. J. (1972). "The visual system and 'neuronal specifcity'." *Nature* **237**, 375–378.

Geren, B. B. & Raskind, J. (1953). "Development of the fine structure of the myelin sheath in sciatic nerves of chick embryos." *Proc. Nat. Acad. Sci. USA* **39**, 880–884.

Hamburger, V. (1934). "The effects of wing bud extirpation on the development of the central nervous system in chick embryos." *J. exp. Zool.* **68**, 449–494.

Hamburger, V. (1939). "Motor and sensory hyperplasis following limb-bud transplantations in chick embryos." *Physiol. Zool.* **12**, 268–284.

Harrison, R. G. (1904). "An experimental study of the relation of the nervous system to the developing musculature in the embryo of the frog." *Amer. J. Anat.* **3**, 197–220.

Harrison, R. G. (1907a). "Experiments in transplanting limbs and their bearing upon the problem of the development of nerves." *J. exp. Zool.* **4**, 239–281.

Harrison, R. G. (1907b). "Observations on the living developing nerve fiber." *Anat. Rec.* **1**, 116–118.

Harrison, R. G. (1910). "The outgrowth of the nerve fiber as a mode of protoplasmic movement." *J. exp. Zool.* **9**, 787–846.

Hope, R. A., Hammond, B. J. & Gaze, M. (1976). "The arrow model: retinotectal specificity and map formation in the goldfish visual system." *Proc. Roy. Soc. Lond. B.* **194**, 447–466.

Hughes, A. F. (1961). "Cell degeneration in the larval ventral horn of *Xenopus laevis* (Daudin)." *J. Embryol. exp. Morph.* **9**, 269–284.

Jacobson, M. (1968a). "Development of neuronal specificity in retinal ganglion cells of *Xenopus*." *Devl. Biol.* **17**, 202–218.

Jacobson, M. (1968b). "Cessation of DNA synthesis in retinal ganglion cells correlated with the time of specification of their central connections." *Devl. Biol.* **17**, 219–232.

Levi-Montalcini, R. (1950). "The origin and development of the visceral system in the spinal cord of the chick embryo." *J. Morphol.* **86**, 253–283.

Prestige, M. C. (1967a). "The control of cell number in the lumbar spinal ganglia during the development of *Xenopus laevis* tadpoles." *J. Embryol. exp. Morph.* **17**, 453–471.

Prestige, M. C. (1967b). "The control of cell number in the lumbar ventral horns during the development of *Xenopus laevis* tadpoles." *J. Embryol. exp. Morph.* **18**, 359–387.

Prestige, M. C. & Willshaw, D. J. (1976). "On a role for competition in the formation of patterned neural connexions." *Proc. Roy. Soc. Lond. B.* **190**, 77–98.

Rakic, P. & Sidman, R. L. (1973). "Sequence of developmental abnormalities leading to granule cell deficit in cerebellar cortex of weaver mutant mice." *J. Comp. Neurol.* **152**, 103–132.

Sidman, R. L. (1974). "Contact interaction among developing mammalian brain cells." *The Cell Surface in Development* (Moscona, A. A., Ed.) Wiley, New York.

Simpson, S. A. & Young, J. Z. (1945). "Regeneration of fibre diameter after cross-unions of visceral and somatic nerves." *J. Anat. Lond.* **79**, 48–65.

Sotelo, C. & Changeux, J. P. (1974). "Staggerer mutant." *Brain Res.* **67**, 519–526.

Sperry, R. W. (1963). "Chemoaffinity in the orderly growth of nerve fiber patterns and connections." *Proc. Nat. Acad. Sci. USA* **50**, 703–710.

Weiss, P. (1934). "*In vitro* experiments on the factors determining the course of the outgrowing nerve fibre." *J. exp. Zool.* **68**, 393–448.

Willshaw, D. J. & von der Malsburg, C. (1976). "How patterned neural connections can be set up by self-organization." *Proc. Roy. Soc. Lond. B.* **194**, 431–445.

INDEX

Acetabularia 3–5, **4**
activation of egg 19–20, 26–27
adrenomedulla **90**, 91
allantois 196–198, **197**, 202, 203
allometric growth **8**, 176
amnion 196–198, **197**, **200**, **202**
amniotes 196
amphibians
  gastrulation 65, **66**, 67–69
  lens induction 104–105, **106**
  lens regeneration 96–**97**
  limb development 128–131, **130**
  limb regeneration 149, **150**
  neurulation 77–81
  nuclear transplantation 32
  optic-nerve regeneration 218
  primary induction 110, **111**
Amphioxus, gastrulation 65
androgens 168–169
annual fish 51, **52**
apical ectodermal ridge 133, **134**, 136–139, **138**
archenteron **64**, 65, **66**, 68, 74, 105, **106**, 112
area opaca **70**, 71, 74
area pellucida **70**, 71
Ascaris, chromosome diminution 16–17, 32
automata theory 142
autonomic nervous system **90**
avian embryo
  cerebellum 215–216
  cleavage 30, **31**, 69
  gastrulation 69–74, **70**, **72**, **73**
  limb development 131–140, **132**, **134**, **138**, 147–149, **152**, 153–154, 163–165, **164**
  motor neurones in spinal cord **210**, 211–212
  mutants:
    talpid[3] **152**, 153–154, 162–165, **164**
    wingless 157
  neurulation **80, 81,** 82, 86
axon, nerve 104, 208–209, 212, 217

behavioural defects
  Drosophila 181
  mouse **216**, 217–218
blastocoele 28
  amphibian **66**
  avian embryo **31**, 74
  mouse **31**, 199–201, **200**
blastocyst **31**, 199, **200**
blastoderm
  avian embryo **70**, 71, **80**, **81**, 82, 85, 198
  Drosophila **16**, 30, 180, **182**
  insects **75**, 76
blastodisc **31**, 71, 74
blastokinesis **75**, 76
blastomeres 28
  amphibians 28, **29**, 36
  frog **29**
  mouse **31**, 93, 199
  sea urchin 39, **40**
blastopore **66**, 67, 74, 110–112, **111**
blastula 28, 63, 67
Bonellia 168
brain 215–223
bristles (trichogens) **178**, 179
BUdR (5-bromodeoxyuridine) 99

cAMP (cyclic adenosine monophosphate) 11–12, 14, 85, 115, 169, 170
canalization of development 93, 156
Cartesian transformations **79**, 81, 155, **156**
cartilage 12–14, **13**, 14, 88, 92, 94–95, 98, 107–108, 131, 139–140, 151, 158
catastrophe theory 151
cations, inducers 115
Cecropia (silk moth) **172**, 173
cell
  adhesion 61, 185
  aggregation 12, **13**, 49, 51, 123, 140, **164**
  communication 83, 85–86

242

INDEX

cell-*continued*.
  death 48, **132**, 133, 135, 212, 218
  fusion and hybridization 34
  migration 17, 89, 159
  movement 48, 51–60, **52**, **57**, **59**, 135, 140, 162–163, 183
  sociology 9, 48, 53
  surface 61, 162–163, 169
cerebellum 215–218, **216**
cGMP (cyclic guanosine monophosphate) 115, 169
chalones 119
chemotaxis 12, 17, 213, 215
chimaeras 204–207, **205**
chorioallantoic membrane **197**, 198
chorion 196–198, **197**, **200**, **202**, 203
chromosome diminution 17
chromosome puffs 174–175
cleavage 28
  avian embryos 30, **31**, 69
  determinate 30, 36, **37**, 43
  frog **29**
  in amphibians 35
  insect embryos 30
  spiral 28, 36, **37**
clonal analysis **178**, 180, 185, **186**, 187
collagen 58, 85, 107, 113, 120, 127, 131, 213
compartments 185–187, **186**, 192–195, **194**
compound eye 6, **7**
contact guidance 89, 213
contact inhibition **54**, 55, 60
cornea 58–60, **59**, 121
cuticle, insect 174, 176, 177, 179, 181

dedifferentiation 95
dermatome **87**, 88
desmosomes **79**, 80, 83
determination 92–102, 139, 141–144, 177, 191, 221
Dictyostelium 9–12, **10**, **12**, 14, 70
differentiation 9, 92–102, 135, 141–142, 162, 221
DMSO (dimethylsulphoxide) 102
DNA 2, 5, 11, 17, 23, 83, 88, 98, 142, 208
dorsal root (sensory) spinal ganglion 87–89, **87**, 210, 214
Drosophila
  bristle patterns 151
  cell markers **178**, 179
  compartments 192–195, **194**
  compound eye 6, **7**
  embryogenesis **16**, 30, **75**, 76, **182**
  genetic assimilation 157

homeotic mutants 193–195, **194**
in vitro cell culture 177
lethal mutants 47, 179
metamorphosis 173
mutants:
  grandchildlessness 15, 47
  Hyperkinetic 181
  multiple wing hairs **178**
  Notch 165
  singed 180
  yellow 180
nuclear transplantation 34
oogenesis **24**, 25
transdetermination 189–191, **190**
wing development 177, **188**, 192–195, **194**

ecdysiotropin 173
ecdysone 170–171, 174–175
EGF (epidermal growth factor) 119–120
egg (ovum) 15, 17–22, **18**, **21**, 34
egg cylinder 201
embryonic shield **70**, 71
endoblast 74
epiblast **31**, **70**, 71, 85
epiboly 67, 71, 198
epigenesis 2
epigenetic landscape 92, **93**, 94, 146, 155–156
epigenetics 2
erythropoiesis 99–102, **101**
Euscelis 43–46, **44**, **46**, **75**, 76, 143
evolution 155–156, 165, 198
extracellular matrix 85, 92–93, **106**, 107, 135
extraembryonic membranes
  amniotes 196–198, **197**
  insects **75**, 76

fate-maps 62, **132**, 137, 180–181, **182**
feathers 107–108, 120–127, **122**
fertilization 19–20, 27
  frog **29**
  mouse **21**
  sea urchin 27
fibroblast **54**, 55, **56**, 58, **59**, 95, 213
fields 128, **130**, 144–145, 192
flask cells **66**, 68, 78
follicle cells **21**, 22–23, **24**, 199
Forficula, polymorphism 176
freemartin 168
Fucus
  polarity in eggs 34

GAG (glycosaminoglycans) 85, 113
gametes 15–26, **18**

INDEX

gap junctions 83–85, **84**
gastrulation 51, 53, 63–73
  amphibians 65–69
  avian embryos **80, 81**, 82
  mouse 199–201, **200**
  sea urchin 63–65, **64**
gene activity, control of 5, 32, 165, 174–175, 193–195, **194**
gene amplification 100
genetic assimilation 156, 157
genetic mosaics
  Drosophila 180, 181
  mouse 203–207, **205**
genotype 155
germ band **75**, 76
germ plasm 15
glial cell 208, **216**, 217–218
gonads 17
gradients
  insect embryos 43–45, **44, 46**
  insect epidermis 183–185, **184**
  limb bud 146–149, 153–154
  sea urchin 41
grey crescent **29**, 35, 38, 67
growth cone 213, **214**, 221

haemopoiesis 99–102, **101**
hairs 120–121, 123, 127, 204, **205**, 206–207
Hensen's node **80, 81**, 82
hormones 93–94, 115, 166–175
Hydra 145
Hydroides, fertilization 20
hypoblast 74

Ilyanassa, polar lobe 36, **37**
imaginal discs 6, **7, 188**, 189
immune system 159
implantation 199–201, **200**
inductive interactions 93. 103–115, 120–121, 123, **124**, 143, 170
informosomes 23, 25
inner cell mass **31**, 93, 199–201, **200**
insects 5, 176–195
  fertilization 20
  morphogenetic movements 74–76, **75**
  nurse cells 23
  social 6, 25–26, 176
instructive interactions 105, 108
involution 67–68

juvenile hormone 171

keratocyte (keratinocyte) 116–119, **118**

lampbrush chromosomes 25
Lebistes 168
lens 96, **97**, 104–105, **106**, 113, 144
leukopoiesis **101**
limb bud 128–140, 147–149, 152–154
limb disc 128–129, **130**, 131
Lymnaea 23, 30, 104

mammals 196–207
  skin development 116–120
mammary gland 94, 108, 166–168, **167**
maternal effects 47
meiosis 17
melanoblasts 89, 98, 158–159, 206–207
melanocytes 98, 104, 116, 158–159, 204, **205**, 206
metabolic co-operation 83
metamorphosis 1
  amphibians 78
  ascidians 78
  insects 6, 170–175, **172**
  sea urchins **64**, 65
metanephric kidney 108, **109**, 110, 112–113
metaplasia 96, **97**, 121
microfilaments
  amphibian neurulation 78, **79**, 80–81
  ascidian metamorphosis 78
  avian neurulation 86
microtubules
  amphibian neurulation 78, **79**, 81
  avian neurulation 86
  insect oogenesis 35
  sea urchin gastrulation 65
  sperm 17, 19
mitosis 48, 71, 99, 117, 119, 135, 137, 180, 222
modulation 95
morphogen 11, 85, 93, 145, 151–153, **152**
morphogenetic movements 62–76, **64**, **66**, **70**, **72, 73, 75**
morula 28, **31**, 199, 204
mosaic development 36, 41, 62, 93, 103, 142
mouse
  cerebellum **216**, 217–218
  cleavage **31**, 199
  determination in early embryo 93
  embryogenesis **31**, 198–203, **200, 202**
  erythropoiesis 100–102, **101**
  gametogenesis and fertilization **21**
  mutants:
    congenital hydrocephalus 158
    nude 159
    Steel 158–159
    T-locus 160–162, **161**
    Tabby 204, **205**, 207

mutants-*continued.*
  testicular feminization 166–168, **167**
  weaver **216,** 217–218
  white spotting 158–159
mouse/chick skin interactions 121, 123
myotome **87,** 88

neural crest 58, **79,** 88–91, **90,** 104–105, 116, 206, 209, **210, 211**
neural plate 77, 81, 86, 88, **106,** 211
neural tube 77, **79, 80,** 81, **87, 114,** 208–209, **210,** 211, **214,** 215
neuroblast 91, 208, 211, 215, 217
neuronal specificity 218–223
neurone 211–215, **214,** 222
neurotransmitters 91, 115, 170
neurula **66,** 106
neurulation 77–86
  amphibians 77–81, **79,** 112
  avian embryos **80,** 81, 82, 86
  mouse 203
notochord **79, 81,** 82
nuclear transplantation 3, 32–34, **33**
nucleocytoplasmic relationships 5, 32, 34
nurse cells 22, 23–25, **24**

oestrogens 168–169
ommatidia 6, **7**
Oncopeltus
  cuticle pattern 181, **184**
  segmental compartments 185, **186,** 187
oogenesis **18,** 20, 22–23, 35
  amphibians 47
  Drosophila **24**
  mouse **21**

parasympathetic ganglia **90,** 91
parthenogenesis 26, 73
Patella, mosaic development 36
paternal genes 47
pattern 141–154
  feather development 123–127
  insect cuticle 181
  limb development 146–154
  mammalian coat pigment 204–207, **205**
  nerve connections 222–223
periblast **31,** 71, 74
permissive interactions 105, 107
phenotype 94–95, 98, 155
pheromones (social hormones) 176
placenta 196–198, **197**
pleiotropy 157–159, 163, 165, 179
pluteus larva 39–41, **40, 64,** 65

polar body 21, 22, 27, **29, 31,** 35
polar co-ordinates 149, **150,** 192
polar lobe 36, **37**
polarity
  amphibian eggs 35
  Fucus eggs 34
  insect epidermis 183–185, **184**
  optic tectum **220,** 221
  vertebrate limb 129–131, **130**
pole plasm 15, **16,** 25, 35
polyclone 187, 193, 195
polyembryony 73
polymorphism
  insects 6, **8,** 25–26, 176
polyspermy 20, 27
polytene chromosomes 174–175
positional information 104, 112, **138,** 139, 144–149, **146, 148,** 183, 192, 221
preformationist theory 2
prepattern 151
primary induction 103, 112, **114,** 143–144
primitive streak
  avian embryos **70,** 71–74, **72, 73, 80, 81,** 82
  mouse 160, **161, 200, 202,** 203
primordial germ cells 15, 17, **18,** 22, 35, 74, 93
progress zone 137, **138,** 145
pseudogastrulation 69
Purkinje cell 215–218, **216**

quail cells, in chick embryo **90,** 91, 140
quantal mitosis 99

regeneration 1, 96, **97,** 149, **150,** 218
regulation **29,** 38, **40,** 41, 73, 93, 104, 144
Reichert's membrane **200,** 201–203, **202**
retino-tectal nerve connections 215, 218–223, **220**
Rhodnius
  cuticle pattern 181, 183, **184**
  metamorphosis 171–174, **172**
RNA 47, 83, 113, 169, 175
  activator 165
  messenger (mRNA) 3, 5, 11, 23, 25, 45, 47, 92
  ribosomal (rRNA) 23–25, **24**
  transfer (tRNA) 23

Schwann cells 89, 209, 211
sclerotome **87,** 88
sea urchin
  gastrulation 63–65, **64**
  regulation 39–41, **40, 42,** 143
self-assembly of molecules **106,** 113, 120
sex determination 17, 26, 168–169

skin 95, 108, 116–127
slime moulds 9, 11, **12**, 14, 151
Smittia, double abdomen in 45, **46**
somite formation **80**, **81**, 82, **86**, **87**, 88, 151, 203
sperm 17–20, **18**, **21**, 26
sperm path **29**, 35
spermatogenesis 17–19, **18**
 mouse **21**
spinal cord **210**, 211
sponges 48–49, **50**
stem cells 95, 99–100, 119, 162
Styela, egg cytoplasm 35
sympathetic ganglion **210**
systems-matching 222–223

temperature-sensitive mutants 179
teratology 1
teratomas 162
testosterone 166
thymus 159
thyroid hormones 170
tight junctions 83, 199
transcription 3, 5, 47, 169, 175
transdetermination 96, 98, 143, 189–191, **190**
transdifferentiation 96–98, **97**
translation 3, 5
trophectoderm **31**, 199, **200**
trophoblast 93, 199–201, **200**

veliger larva 36, **37**
vitamin A 95, 119
vitelline membrane **16**, 23, 27, **70**, 71

Wolffian regeneration 96, **97**
wound-healing 1, 174

X-chromosome inactivation 204, **205**
Xenopus
 anucleolate mutant 32
 cleavage 28
 gastrulation 69
 germ plasm 15
 motor neurones in spinal cord 212
 neurulation 78, **79**
 nuclear transplantation 32, **33**
 retino-tectal connections 219–222, **220**
 RNA synthesis in oogenesis 25
 sex determination 168

yolk 22, 30, 198–199
yolk sac 196–198, **197**, **200**, 201, **202**

zebra 207
zona pellucida **31**, 199
zone of polarizing activity 131, 139, 147–149, **148**